To Karen, my helper, consultant, and inspiration.

She makes my heart go whoosh *and* boom.

Contents

Introduction

Since my first book, *Backyard Ballistics*, was published, a couple of things have happened. First, the number of books in my library devoted to researching things that go *whoosh*, *boom*, and *splat* has grown enormously. Concomitantly, the bookshelf holding biographies of inventors has also grown quickly.

Many of the scientists and practical philosophers associated with this field of study are pretty well known. For example, books about Alfred Nobel (the inventor of dynamite), Robert Goddard (the father of liquid fuel rocketry), and Archimedes (the greatest of the ancient Greek scientists and discoverer of the principle of buoyancy) are easy to find. But many more scientific innovators are not as well known as they should be. In this book you will meet several scientists and historical figures who perhaps should be more famous, considering how adept they were at making things go boom and how the events that followed their work changed the course of history.

The names Urban of Hungary, Federico Gianibelli, and Waldo Semon are just about forgotten, and more important, world-changing scientists like Alexander von Humboldt and Count Rumford are familiar to only a few. These men were able experimenters and terrific scientists who shaped wood, cut metal, mixed chemicals, and occasionally got into trouble. But they also did great things.

The other thing that's happened since the publication of *Backyard Ballistics* is that people often ask me, "Are you the fellow who wrote that book about blowing things up?"

Well, golly—no, I'm not.

I never really think of *Backyard Ballistics* as being about blowing things up. It is not a cookbook for anarchists, and there are no recipes in it for making bombs. The projects in *Backyard Ballistics* do result in things going boom, at least occasionally. But is that a good thing? Is there anything inherently good and noble about experimentation and fabrication, when the desired result is nothing more than a really good bang? I myself wasn't sure until I recalled something a friend told me several years ago.

This particular friend, Jim, is a tall, laconic electrical engineer, a product of the northern plains. He grew up in a rural area in northwestern Minnesota, around the towns of Red Lake, Fertile, and Climax. As a boy, he spent his free time taking apart farm machinery and gadgets and then putting them back together. When he was old enough, he joined the air force, hoping to see the world. In the irony of military life, Jim actually wound up seeing more of the northern prairie, since he was assigned to a program targeting instructions for the nuclear missiles that were then hidden underneath North Dakota.

During the 1960s Jim tinkered on the hundreds of Minuteman intercontinental ballistic missiles buried in pop-top tombs under the Dakota winter wheat fields. At the time of his enlistment there were so many rockets up there that if North Dakota had seceded from the Union, it would have become the world's second-largest nuclear power.

Jim's work involved targeting nuclear missiles by straddling the nose cone at the tip of the rocket and shooting the coordinates of visible landmarks through a transitlike device called a theodolite. Then he fed a metal punch tape into the tape-reader fronting the missile guidance system in the communications bay of the rocket. The tape provided the exact coordinates of that week's top Cold War target—say, the Soviet Strategic Rocket Force bases at Tatischevo, or the Northern Fleet Command bunker at Severomorsk. This told the sleeping Minuteman exactly where it should head, should the go-code arrive from the Pentagon to wake it up.

From this background of agrarian and military tinkering, Jim took the logical step of enrolling in college after his discharge. Because he enjoys tinkering, he earned his degree in engineering.

The love of tinkering and a concomitant affection for making things go whoosh, or zoom, or even boom, should be the identification badge by which modern engineers are recognized. "The greater a person's tinkering skills," Jim told me, "the greater the engineer."

In Jim's youth, way up there on the rural northern plains, there weren't many things to fill the leisure time of teenage boys. Two of the most popular pastimes were drinking and tinkering. While some misspent their youth throwing back beers in ice-fishing shacks on frozen lakes, others tinkered. To certain boys, the long, icy nights and the scarcity of outside stimuli weren't privations. They were gifts of time: the cold, clear, and quiet air of the prairie served as a lens to focus their attention. Myriad projects invariably pose themselves to a plains state tinkerer with time on his hands.

Jim observed that those rural and small-town boys (they were mostly boys back then) in the mid-twentieth century who filled their free time with tinkering got good at it. Many who left the farmlands followed the very same path to study engineering as my friend Jim did. They had a practical basis for their knowledge and study, rooted in long hours of advanced tinkering and, occasionally, blowing something up—not violently, not dangerously, but on purpose and with care. They became good, even great engineers.

These engineers looked at themselves not just as technicians or professional designers but as engineer-artists whose creations were both mechanical art and extensions of themselves.

To predict the adult engineering success of a youth, look to his or her propensity for making things go whoosh, boom, or splat. Thomas Edison's childhood chemistry experiments, performed in the baggage car of the train on which he worked, resulted in the accidental destruction of the railroad car. B. F. Skinner, the great social scientist, was proud of the steam cannons and catapults he built as a youngster. Homer Hickam, the NASA engineer who wrote the best-selling book *Rocket Boys*, predictably enough burned down part of his yard with primitive rocket experiments. A certain Thomas Shelton, of whom we'll learn more, accidentally demolished his parents' kitchen with an experiment gone wrong. And a young engineer named Jack Welch burned down one of his employer's factories, yet managed to retain his job. He went on to become the CEO of General Electric.

Making things go *whoosh*, *boom*, and *splat* is a rite of passage for the intellectually curious. Most of the time such events don't happen on purpose, but happen they do. This book does not advocate reckless experimentation. Danger and recklessness are the antithesis of good experimental technique. I still remember one particularly pithy comment that my friend Jim made to me many years ago. "Science isn't just about blowing things up," he told me. "Rather, it's about blowing things up and knowing how you did it."

So with this in mind, read chapter 1 carefully.

Safety Guidelines

The projects described in the following pages have been designed with safety foremost in mind. But as you try them, something unexpected may still possibly occur. It is important that you understand that neither the author, the publisher, nor the bookseller can or will guarantee your safety. When you try the projects described here, you do so at your own risk.

Some of the projects included in this book have been popular for many years, while others are new. Unfortunately, there have been rare instances of damage to both property and person when something went wrong. The likelihood of such an occurrence is remote so long you follow the directions, but still, things can go wrong. Always use good common sense, and remember that you carry out all experiments and projects at your own risk.

Also, be aware that every city, town, and municipality has its own rules and regulations, some of which may apply to these projects. Further, local authorities have wide latitude to interpret the law. Therefore you should take time to understand the rules, regulations, and laws of the area in which you plan to carry out these projects. A check with local law enforcement can tell you whether a project is suitable for your area. If not, there are plenty of other places where all of the projects here can be undertaken safely and legally. Be sure to check first!

GROUND RULES

When you were a kid, people told you not to play with matches for a good reason: They can be dangerous. Always follow the instructions closely. If you don't, making the projects in this book has the potential to harm you and your stuff. Don't make changes to the materials or construction techniques.

These are your general safety rules. Each chapter also provides safety instructions specific to its project.

1. The projects described here run the gamut from simple to complex. Many, if not most, of them are designed for adults or, at a minimum, should be closely supervised by adults. Take note: Some of the projects involve the use of matches, volatile materials, and projectiles. Adult supervision is absolutely mandatory for all such projects.

2. Read the entire project description carefully before you begin. Make sure you understand what the project is about and what you are trying to accomplish. If something is unclear, reread the directions until you fully comprehend the entire project.

3. Keep people away from the firing zone in front of and to the sides of rockets, mortars, cannons, and the like. Use care when transporting,

aiming, and firing, and always be aware of where the device is pointing.

4. Don't make substitutions for the specific liquids and aerosols indicated. Stay away (far away) from gasoline, kerosene, alcohol, and other flammable liquids. Few things are as dangerous as flammable liquids and aerosols. They can and do explode, and the consequences can be disastrous.

5. Use only the quantities of fluid listed in the project's materials list. Don't use more propellant than specified.

6. Don't substitute materials or alter construction techniques. If the directions say to cure a joint overnight, then cure it overnight. Don't take shortcuts.

7. Read and obey all label directions for PVC primer, cement, and any and all other chemicals. Pay close attention to directions regarding ventilation.

8. Remove and safely store all cans and bottles containing naphtha, hair spray, or any other flammable substance prior to launching the project. A good rule of thumb is to maintain a hazard-free radius of at least fifty feet around the area in which you plan to work.

9. Clear the area in which you will undertake the projects of all items that can be damaged by projectiles, flying objects, and so forth.

10. Be aware of pinching hazards on projects with mechanical arms or other moving parts. Keep fingers and toes clear of pinch points.

11. Wear protective eyewear when indicated. Some experiments call for hearing protection, blast shields, gloves, and so forth. Always use them.

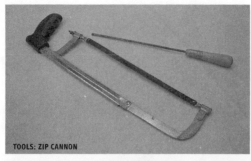

TOOLS: ZIP CANNON

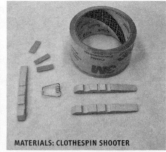

MATERIALS: CLOTHESPIN SHOOTER

TOOLS: VORTEX CANNON

TOOLS: MECHANICAL TOE

TOOLS: ARCHITRONITO

MATERIALS: O-RING FOR T-SHIRT CANNON

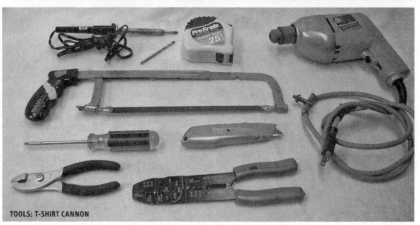

TOOLS: T-SHIRT CANNON

CHAPTER 2

Materials

Several of the projects in this book involve cutting and joining PVC pipe. This section tells you how to make safe and secure joints.

First, you should be aware that at least four types of plastic pipe and plastic pipe joints are available: PVC, CPVC, ABS, and PB. The letters are abbreviations for the type of plastic polymer that composes the pipe. Pressure-rated, schedule 40 PVC pipe and pipe fittings are made of white polyvinyl chloride. This is the type of pipe and pipe joint recommended for most of these projects.

WORKING WITH PVC

PVC pipe is easily cut with a regular fine-bladed handsaw. It is important that you make all cuts as close to 90 degrees to the centerline of the pipe as possible. That way you won't leave any interior gaps, which will weaken the joint.

Before you apply any cement, you may want to *dry-fit* a pipe into a joint just to see how it fits. Sometimes dry-fitted pipes and joint fittings stick together so tightly, it is hard to get them apart. If that happens, carefully whack the fitting loose with a wooden block.

Joining and gluing PVC pipe is technically called *solvent-welding.* The solvent melts the plastic so that when you push the pipe and the pipe fitting together, the two parts fuse as the solvent evaporates. Each type of plastic pipe has its own special solvent. Some solvents are advertised to work on several types of plastic, but it is recommended that you use the solvent that is meant specifically for the type of plastic you're working with.

Solvent works only on *clean* surfaces—that means no dirt, no grease, and no moisture. Wipe the inside of the fitting and the outside of the pipe with a clean cloth. Then apply PVC primer (called purple primer) to the ends.

Next, coat the surfaces that you want to join with a liberal amount of PVC solvent. (PVC solvent should be used only in a well-ventilated area.) Push the pipe into the pipe fitting quickly, and give it a one-quarter turn as you seat it. Hold it tight for about fifteen seconds, and voilà! The joint is set. Be sure to observe the cure time shown on the PVC cement bottle's directions. PVC joints must cure overnight to reach full strength.

Take note that PVC pipe's material characteristics change as the temperature changes. At temperatures below 55 degrees and above 95 degrees Fahrenheit, PVC's mechanical properties are impaired, and devices constructed from PVC should not be used.

Be aware that PVC solvent fumes are extremely flammable. Keep ignition sources away from fresh PVC constructions for several hours or until the solvent-welds are cured and the fumes are gone.

OBTAINING MATERIALS AND TOOLS

Most of the materials and tools used for the projects in this book are available at larger hardware stores and lumberyards. A few projects require some basic electrical supplies such as switches, batteries, and wiring, which are available at stores such as Radio Shack or any electrical supply store.

If you cannot find a local parts supplier, several large industrial supply houses may be able to provide the parts you need. Supply companies such as McMaster-Carr Supply Company (www.mcmaster.com) have large inventories and well-organized websites.

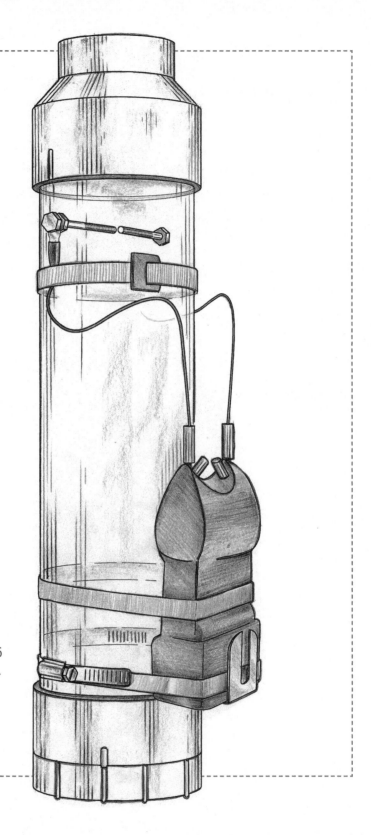

The Night Lighter 36
high-voltage ignitor
and combustion
chamber

CHAPTER 3

Playing with Fire

Prior to 399 B.C. there was no such thing as artillery. Ancient warfare consisted of hand-to-hand combat—club against club or sword against sword. In addition, warriors used thrown weapons like spears and stones, which were usually hurled from slings. Finally, they shot arrows from stretched bows.

In 399 B.C. military leaders in the Greek city of Syracuse came up with ideas for larger, semistationary weapons that could wield much more power than could a single warrior's arm. These devices used animal parts, such as horns or neck tendons or hair, that had been fashioned into springs. By stretching these springs, a man (or even better, a group of men) could load the device with comparatively huge amounts of potential energy. These weapons were the earliest catapults and represented the first type of artillery.

From 399 B.C. until A.D. 1331 all military artillery was powered by human muscle. During this 1,730-year stretch, artillery took the form of siege weapons called onagers, trebuchets, ballistas, mangonels, petraries, and spring engines—all of which we now refer to by the general term *catapult*.

Catapults work by suddenly releasing energy that has been loaded into the device by work from the operators' muscle. In ancient and medieval times catapult operators would tighten a rope spring, bend back an enormous bow made from wood and animal horn, or raise a great weight high off the ground. When a catapult is fired, a spring releases or a counterweight falls and, through a mechanism or lever, quickly and efficiently transfers its stored energy to the projectile.

Catapulting a bigger stone naturally requires a larger counterweight or a thicker composite bow. By the end of the thirteenth century, catapults were getting about as big as they could get. The law of diminishing returns was taking over; making catapults any larger and more powerful would push the materials from which they were made to their mechanical limits. Moving and erecting the largest catapults already required gigantic siege trains and armies of men. An entire forest had to be cut to provide wood for such a device. Given the deforestation throughout Europe at the time, building such giant catapults became an ecological disaster as well as a logistical nightmare. Most important to kings and princes, they simply became an unacceptably large drain upon royal treasuries.

Such was the state of artillery warfare throughout the Middle Ages. But then a new discovery, brought by explorers and traders returning from the Far East, changed the world. The Chinese called this discovery the "fire drug." We know it today as gunpowder.

Fire-based, or incendiary, projectiles were used in warfare dating back to the ancient Greek and Chinese cultures. Ancient accounts tell of baskets of flaming oil being hurled at enemies, arrows tipped with heads of burn-

Ancient Greek mechanical arrow shooter, or ballista

Medieval trebuchet hurling a flaming barrel

ing straw, and even the Fire Ox, a Chinese innovation that is just what it sounds like—an ox with a barrel of flaming goo strapped to its hindquarters and sent stampeding toward enemy lines. But all of these were weapons of conflagration, not explosion. They spread fire, but they did not blow up with great and destructive energy.

Around A.D. 1000, the Sung dynasty in China began experimenting with the "fire drug," a compound of charcoal, sulfur, and a white powder that leached to the surface of the ground in certain areas. The white powder was potassium nitrate, which is often referred to as saltpeter.

By themselves, charcoal and sulfur do a pretty good job of burning. They release energy when ignited, but they do it rather slowly. The

Chinese discovered that adding saltpeter to a mixture of sulfur and charcoal caused the compound to burn far more rapidly. And if the ratio of charcoal, sulfur, and saltpeter was exactly right, the compound would explode when heated in an enclosed space.

The Chinese military quickly made use of this discovery. They placed the fire drug mixture in various types of projectiles and bombs, giving them elaborate names such as "Bandit-Burning Vision-Confusing Magic Fire Ball," "Match-for-Ten-Thousand-Enemies Bomb," and "Heaven-Shaking Thunder Crash Bomb."

The Chinese had no idea why this mixture of chemicals was so powerful, but nevertheless it was. Scientists now know that gunpowder is a fuel-oxidizer chemical combination. Things burn because they react with oxygen, and the more oxygen, the better things burn. Saltpeter is an efficient oxidizer. When it is burned, a chemical reaction releases great numbers of oxygen molecules that combine with the burning sulfur-charcoal fuel. The result is an ultravigorous energy release. When this reaction occurs in a closed container, an explosion results, creating a weapon worthy of the appellation "The Bone Burning and Bruising Magic Oil and Smoke Bomb."

Ancient chemists tried different proportions of charcoal, sulfur, and saltpeter and found that minute proportional changes resulted in much improved results. A bit more saltpeter and a bit less sulfur here and there would substantially increase the power of the chemical compound. Over time the experimenters found that the most effective proportion was about 75 percent saltpeter, 15 percent charcoal, and 10 percent sulfur.

Once the correct proportions of the "devil's distillate"—as gunpowder was called—were determined, military men began to consider how to use it. One of the most straightforward applications was as a propellant for cannonballs.

The first semitrustworthy report of the use of a gunpowder cannon (or at least primitive cannonlike technology such as a serpentine or culverin) in battle occurred in 1324. In the fourteenth century the city of Metz had become one of the largest in eastern France because of its position on the intersection of important trade routes.

Rivaling some of the great cities of Flanders, Italy, and Germany, Metz was a tempting target for power-hungry noblemen. In 1324 the Dukes of Luxembourg and Lorraine attempted to take the city by force. According to the old chronicles, one of the leaders of the siege, William de Verey, sailed into the town onboard a flatboat, loaded not with horsemen or catapults but with new and secret weapons.

Up the River Moselle went the battle barge. On its deck were things that had not been seen in warfare previously but that would be used in virtually all battles from that time onward.

An account written in 1838 by French historian J. Huguenin in *Les Chroniques de la Ville de Metz* relates that on this barge were "*serpentines et canons . . . et tirant plesieurs coptz de' artillerie*" (serpentines, cannons, and many other types of shooting artillery). According to this town history, it was here, at the walls of Metz, that European besiegers first packed the new explosive called gunpowder into iron vases.

The chroniclers of the battle don't tell us what effect, if any, the primitive rock-firing cannons had on the outcome of the siege. Most likely, despite the presence of gunpowder cannons, the battle was won by the tried and true methods of medieval siege tactics: escalade (scaling walls with ladders), sapping (digging tunnels), frontal assaults, and subterfuge. But in a short time artillery improved to the point where such medieval siege tactics were no longer relevant. Modern warfare probably got its start right here, on a river barge filled with crude cannons that literally could not hit the broad side of a barn.

LEVELING YOUR WEAPON

As must have been apparent to any fourteenth-century artilleryman in the vanguard of an assault, a round stone or iron ball doesn't travel as far or as accurately as a bullet-shaped projectile. Why? A bullet's cylindrical body and pointed head cut down on air friction—that is, aerodynamic drag—as it flies through the air. And the cylindrical body is designed to spin as it travels up the helical groove of a rifle barrel. The spinning stabilizes the bullet as it's in flight, allowing it to travel farther and straighter.

Bullet-shaped projectiles are a relatively new innovation. Until the mid-1800s firearms were mostly smooth-bored, muzzle-loaded weapons, and most projectiles were round.

If the musketeer put enough gunpowder in the charge, he could impart a reasonable muzzle velocity when the projectile left the barrel. But air resistance acting on the round shape quickly slowed it down. Contrary to intuition, spheres, it turns out, are among the worst possible shapes in terms of creating drag. And cylinders aren't much better. The problem with spheres and cylinders is that the wake they leave behind as they fly through the air is huge in proportion to their cross-sectional area. Big wakes cause big drags.

In the late 1980s Austrian researchers conducted a series of tests on the effect of air friction on bullets for a group of

It's easy to see why Turkish cannon-balls weren't very aerodynamic.

sixteenth-through-eighteenth-century muskets. What they found was surprising. The tests showed that when a spherical bullet is first shot, it decelerates at a rate of about 2.5 meters per second for every meter of distance it travels. Modern bullet-shaped bullets lose only about 0.9 meters per second in velocity over similar distances. In other words, spherical bullets slow down three times as fast as bullet-shaped bullets. After one hundred meters, a musket-fired ball has only half as much kinetic energy as when it first leaves the muzzle. Modern rifle-fired ammunition maintains 90 percent of its energy at the same distance.

In addition to velocity loss, round shot is hard to aim accurately. Instead of flying straight and true toward the target, a round bullet tends to slice and hook like a golf ball. This effect is caused by an unintended spin that is imparted to the bullet around an arbitrary axis. The spin changes the air pressure on one side or the other of the sphere and sends the projectile off course. This phenomenon is called the Magnus effect, named after Gustav Magnus, who first described it in 1852. The veering of projectiles caused by the Magnus effect made it nearly impossible for a musketeer to hit a target.

So as late as the Battle of Waterloo in 1815, soldiers never talked about "aiming" their weapons. Instead, they merely "leveled" them.

HIGH VOLTAGE AND NIKOLA TESLA

Our first project combines mechanical and electrical principles to produce a unique and powerful hand-held spud gun. The Night Lighter 36 is a combustion-based cannon capable of shooting a potato from one end of a football field to the other. Its power and efficiency result from a high-voltage spark.

The previous section provided some history on the mechanical background of this project. This section looks at its history from an electrical engineering perspective, in particular the connection to scientist Nikola Tesla, who brought high voltage into daily life in a meaningful way.

Nikola Tesla was born in 1856 in Smiljan, Croatia. As a child, all who knew him realized that he was in possession of incredible intellect but was simultaneously plagued by myriad mental phobias and compulsions.

Tesla attended several schools and colleges, most notably the University of Prague, until he was banished for excessive gambling and "leading an irregular life." While in Prague, Tesla started work on what would be his most important contribution to science: alternating current, or AC.

Prior to Tesla's work, electrical-power-generation systems were limited to producing low voltages and high currents. Power companies had to either use very expensive, heavy copper transmission cables (to minimize wire resistance) or build a DC generating station every mile or two. This low-voltage method, known as DC power distribution, was favored by the enormously rich and powerful American businessman Thomas Edison.

But high voltage, Tesla felt, was the key to the cheap and ubiquitous distribution of power, and using alternating current was the only way to achieve such distribution. He found a champion in a wealthy businessman named George Westinghouse. A brilliant and far-sighted entrepreneur, Westinghouse had built a giant business enterprise from nothing, making his fortune by manufacturing air brake systems for trains.

The new Westinghouse Electric Company was well financed, well managed, and most important, willing to base its entire business plan on Tesla's idea of AC power, which in almost every way was superior to DC. But Edison would not give up. Even if AC power was exponentially more efficient to transport and cheaper to make, it did have liabilities, and these he would exploit with a Madison Avenue–style public relations appeal to the public.

The Edison camp's plan was simply to portray AC power as too dangerous and deadly for use in homes and businesses. The PR campaign went something like this: AC power uses high voltage to transport power into people's homes. In fact, high voltages in and of themselves make the whole concept fly. But, said Edison's PR minions, high voltages are dangerous. If you accidentally touch high voltage, you can die. Today every room in every home and business has several 110-volt AC outlets, but in 1883 the idea of loading up your house with portals to such deadly currents was nearly unthinkable.

"AC power can kill you and your family," the Edison flacks said, and they backed up their claims with dubious tactics and statistics. Worse, they graphically demonstrated their point by electrocuting (with AC power, of course) scores of animals, barnstorming the country in a ghastly road show. The Edisonians fried scores of stray dogs and cats as well as farm animals (and in Albany, New York, an orangutan). The animal-electrocution cavalcade reached its cynosure with the Coney Island electrocution of a rogue elephant named Topsy.

But despite Topsy and the rest of the PR campaign, Edison's hope of discrediting Westinghouse failed, and soon Westinghouse's AC power was lighting the incandescent lamps of all electrified localities. Tesla's higher-voltage AC model won out, for it was cheaper, faster, and just plain better.

In 1896 Tesla successfully installed a large AC power system at Niagara Falls, New York. The completion of that system ended the "War of the Currents," as this technology competition is popularly called. Soon afterward Edison's General Electric Company capitulated and converted to AC power.

Interestingly, New York City's electric utility company, Consolidated Edison, continued until recently to supply DC current to a few customers

Bad Elephant Killed

Topsy, the ill-tempered Coney Island elephant, was put to death in Luna Park, Coney Island, yesterday afternoon. The execution was witnessed by 1,500 or more curious persons, who went down to the island to see the end of the huge beast, to whom they had fed peanuts and cakes in summers that are gone. In order to make Topsy's execution quick and sure 460 grams of cyanide of potassium were fed to her in carrots. Then a hawser was put around her neck and one end attached to a donkey engine and the other to a post. Next wooden sandals lined with copper were attached to her feet. These electrodes were connected by copper wire with the Edison electric light plant and a current of 6,600 volts was sent through her body. The big beast died without a trumpet or a groan.

Topsy was brought to this country twenty-eight years ago by the Forepaugh Circus, and has been exhibited throughout the United States. She was ten feet high and 19 feet 11 inches in length. Topsy developed a bad temper two years ago and killed two keepers in Texas. Last spring, when the Forepaugh show was in Brooklyn, J. F. Blount, a keeper, tried to feed a lighted cigarette to her. She picked him up with her trunk and dashed him to the ground, killing him instantly.

*From *Commercial Advertiser* [New York], January 5, 1903.

who had specialized needs for it, mostly for old elevators. Only in January 2005 did Con Ed announce that it would cease offering DC service to those fifteen hundred remaining users, ending that chapter of electrical transmission history.

THE NIGHT LIGHTER 36

Author with Night Lighter 36

Spud guns are very popular amateur science projects. Few devices can be made as simply and inexpensively as a spud gun yet provide so much entertainment and fun. This project shows you how to make the Night Lighter 36, a transparent PVC potato cannon with a stun gun ignition system. It can be used both day and night, but the clear PVC tubing provides an excellent view of the internal workings when used in darkness.

The knowhow for fabricating potato cannons is provided in many books and various Internet sites.* There are companies that sell the plans, parts, and completed spud guns over the Internet. Some products feature rather elaborate construction, including rifled barrels and metered fuel systems.

----

*Backyard Ballistics (Chicago Review Press, 2001) includes detailed instructions for making a simple yet dependable one, as well as assorted other projects.

Most spud guns however, are made at home. Having once mastered basic gun construction, the intrepid potato cannoneer often wants to design and assemble more complex and more artistic devices. The word *artistic* is applicable because making spud guns is an art as well as a science. For many, the act of shooting resonates on a primal level, and makers often take great pains to enhance their experience by continuously tinkering with their guns with the ardor of hot-rod builders and model railroaders.

The Night Lighter 36 potato cannon uses inexpensive modern circuitry to produce a stunningly high voltage that ignites a fuel-air mixture in the cannon barrel. It is a relatively simple extension of the basic spud gun concept and is fun to operate and entertaining to behold. A basic, no-frills spud gun can be built for less than twenty-five dollars. The Night Lighter 36 costs more due to the price of the transparent PVC tubes and the stun gun igniter. But with a bit of scrounging, you can often control costs. I built mine for less than fifty dollars by finding odd pieces and leftovers from plastics suppliers. If the cost of the transparent PVC pipe is a problem, you can build this project using regular, nontransparent, schedule 40 PVC pipe.

✷ THE GRANDFATHER OF THE SPUD GUN

The hard white plastic used to build spud guns is called PVC, an abbreviation for polyvinyl chloride. It was invented in 1926 by a scientist named Waldo Semon. At the time Semon was a young research scientist working for the B.F. Goodrich Company in Akron, Ohio.

Semon accidentally discovered PVC while he was attempting to develop synthetic rubber. In World War I Germany's adversaries cut off its supply of natural rubber. In response, the Germans developed rubber substitutes. Their efforts were only partially successful at best—the substitutes

SAFETY

If you want to join the community of spud gun enthusiasts, you must understand a couple of important points.

The first is that spud guns may not be legal in your area. That may seem unfair, considering that in some places it is perfectly legal to walk down Main Street holding a twelve-gauge, but toting a tater gun may get you in trouble. That said, no federal law appears to regulate the construction and use of potato cannons, but state and local laws do. So check with local law enforcement regarding the rules that pertain to your area, then obey them.

Second, spud guns can be dangerous—they can shoot with great power and force. So treat your gun with respect. A few important points to keep in mind:

1. A potato cannon shoots with enough force to cause injury. Its range can exceed 200 yards. Always use extreme care when handling and aiming the device. Wear eye protection.

2. Don't operate a damaged or worn gun. Check the cannon frequently for signs of wear.

3. Don't take construction shortcuts, use inferior materials, or make substitutions in the materials list. But if clear PVC is too expensive for your budget, you may substitute white schedule 40 PVC pipe.

4. PVC's mechanical properties degrade in high and low temperatures. Don't use spud guns in temperatures below 55 or above 95 degrees Fahrenheit.

5. The vapors from PVC cement are flammable. Allow all joints to dry fully before exposing them to ignition sources.

(continued)

6. A misfire is a fairly rare event, but if it does occur, use the utmost caution. Carefully remove the end cap, and ventilate the combustion chamber thoroughly. Never look down the barrel, and never point the barrel at anything you don't want to hit.

7. The stun gun igniter is made from an inexpensive stun gun commonly used for personal protection. Check local laws regarding their possession and use. Stun guns are designed to be high-voltage shocking devices. Touching the high-voltage electricity contacts of a stun gun will hurt. Avoid them!

8. As with many popular activities, things occasionally go wrong. If you follow the directions here carefully, the chance of a bad outcome is remote, but unexpected things can happen: material flaws, bad judgment, pure bad luck. Always use common sense, and take careful note of the following: *Build and use the spud gun at your own risk.*

were not very resilient. Drivers of cars with synthetic rubber wheels would often find that the tires had formed flat spots overnight, resulting in a bumpy ride the next morning. Worse, the synthetic rubber was more expensive than the real thing.

Shortly after the end of World War I, Goodrich determined that a low-cost, high-performance rubber substitute would be a product with enormous economic potential. Since Semon was one of their best scientists, company management instructed him to begin work to find a better-behaved, less expensive man-made rubber.

During his extensive research, Semon came upon the work of earlier scientists who had developed a substance called "oil of Dutch chemists." Another chemist, he found, had created a gas called vinyl chloride from the Dutch oil. Semon felt the stuff had potential, and so he started experimenting with vinyl polymers. Polymers are organic substances with very large molecular chains. After several lab explosions and near misses, his chemical experimentation yielded a hot powder that, when cooled, turned into a flexible but tough gel. This was the first batch of polyvinyl chloride.

Interestingly, Semon took the most pleasure in another of his inventions that was far less heralded—synthetic rubber bubble gum. "It looked just like ordinary gum, except that it would blow these great big bubbles," he said. To his disappointment, it never caught on.

Potato cannons and their relations have been built from pipes made of steel or aluminum, or other types of plastic, or even reinforced fabric tubing. But for cost, strength, and ease of fabrication, PVC is hands down the most popular material. While Dr. Semon might never have made his own spud gun, his invention made them possible.

Using PVC Pipe as a Building Material

A spud gun is largely composed of PVC pipes and fittings that are solvent-welded in place. Solvent-welding must be done properly to prevent leaks and weak spots. This is a good time to review the earlier instructions on working with PVC pipe.

1. Inspect the pipe end and fitting for cracks, dirt, and abrasion. Don't use damaged PVC pipe or fittings.
2. Cut the pipe off squarely to the proper length using a fine-tooth saw or a hacksaw. Remove burrs.

3. Clean the weld surfaces with PVC primer. Apply the primer with a dauber or brush. The primer cleans and softens the PVC. Allow the primer to dry before applying cement.

4. To solvent-weld, brush on a thick coat of PVC solvent cement that is matched to the type of pipe and fitting you are using. Use PVC solvent only. Apply plenty of cement, first to the pipe end, then to the fitting socket. Leave no bare spots.

5. Immediately join the pipe and the fitting full-depth with a slight twist to bring it into correct alignment. A continuous ooze of cement around the fitting indicates that you used enough solvent cement to ensure a leak-free joint. Let the joint dry for several hours before using.

6. Once joined, you can't reposition the pipes or otherwise fix errors. If you accidentally put the wrong fitting on a pipe, you have no choice but to cut it out and start over.

7. Note: PVC solvent vapors are flammable. Let the cannon dry for several hours in a well-ventilated area before use.

GENERAL PVC SOLVENT SAFETY INSTRUCTIONS

1. Avoid breathing PVC solvent cement and primer vapors. Work in a well-ventilated area.

2. Keep all chemicals away from open flames.

3. Read and follow the precautions on the labels.

MAKING THE NIGHT LIGHTER 36

TOOLS

Hacksaw

Power drill

$^{15}/_{64}$-inch drill bit

Sanding drum for power drill

Screwdriver

Pliers

Utility knife

Wire stripper

Safety glasses

MATERIALS

(1) 14-inch-long, 3-inch-diameter transparent (or regular) schedule 40 PVC pipe

(1) 36-inch long, 2-inch-diameter transparent (or regular) schedule 40 PVC pipe

(2) Crimp-on barrel connectors

(1) Stun gun, nominal 100,000-volt output or better

(2) 2-inch-long, $^{1}/_{4}$-inch hex head bolts, full threaded

(4) $^{1}/_{4}$-inch hex nuts

(1) 3-to-2-inch-diameter white PVC reducing fitting

(1) 3-inch white PVC threaded adapter coupling, one side smooth, one side threaded

(1) Can PVC primer

(1) Can PVC cement

(1) 2-foot length of insulated 12-gauge (or heavier) wire

(2) Large crimp-on spade connectors

(2) 15-inch-long (not diameter) hose clamps

(2) Small wire nuts

Electrical tape

(1) Tube of silicone sealant

(1) 3-inch-diameter white PVC threaded end cap

Bag of potatoes

(1) 4-foot long, 1-inch-diameter wooden dowel or broom handle

(1) Can of hydrocarbon containing aerosol spray (hair sprays typically work well)

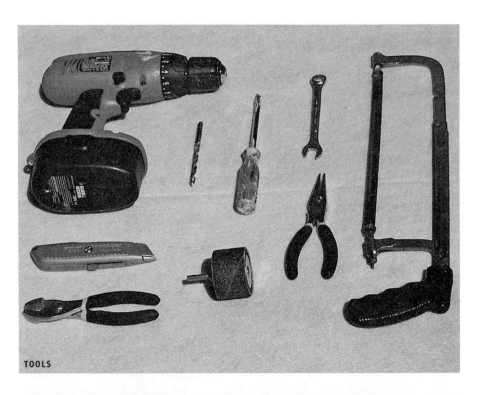

TOOLS

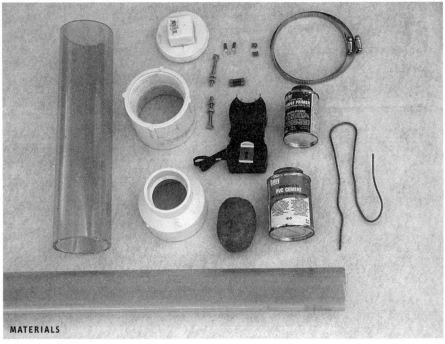

MATERIALS

SOURCES OF MATERIALS

With the exception of the transparent PVC pipe and the stun gun, most of the materials in the list are available at large home and hardware stores.

Transparent PVC, used mainly in the food-processing industry, is available from industrial plastic suppliers. Check the business listings in your telephone directory or search the Internet. Transparent PVC pipe is expensive when purchased at full price, but you may be able to purchase remnants and scrap at greatly reduced prices from a supplier or food-processing company.

Stun guns are widely available from suppliers on the Internet. A search for the word *stun gun* will return a large number of vendors. Mail-order firms, such as Cabela's (www.cabelas.com), often sell them for less than twenty dollars.

MATERIAL PREPARATION

Step 1: Cut the PVC

Measure and mark a cutting line 14 inches from one end of the 3-inch-diameter PVC pipe. Use the hacksaw to cleanly and squarely cut the pipe.

Now, measure and mark a cutting line 36 inches from one end of the 2-inch-diameter PVC pipe. This will be the cannon's barrel.

Step 2: Fillet the edge of the barrel

Use a file and sanding drum attachment to taper one end of the long 2-inch-diameter pipe section so it forms a sharp edge. A clean sharp edge is important since it should cut the perfect-sized potato plug projectile as you ram the potato into the muzzle of the gun.

Step 3: Drill electrode holes

Drill a hole 4 inches from one end of the 3-inch-diameter pipe. The hole should be $^{15}/_{64}$ inches in diameter. Drill a second hole of the same size, 4 inches from the same end, but at the spot that is exactly diametrically opposed to the first hole.

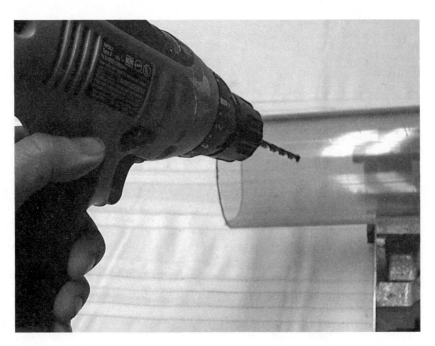

Step 4: Modify the barrel connectors

Using a sharp utility knife, remove excess insulation from the barrel connectors. Now, with the stun gun turned off (temporarily remove the battery if you wish), place the connectors over the stun gun electrode to test the fit. Be sure to use the twin electrodes that point forward rather than toward each other.

NOTE: Depending on the make and model of the stun gun you use, you may need to modify these directions in order to connect the ignition wire to the stun gun electrode. Other types of connectors such as wire nuts and soldered connections may be used if necessary.

CONSTRUCTION

Step 1: Install the electrodes

Screw in the 2-inch-long hex head bolts with two hex nuts attached on each (nuts go outside the barrel) into the holes in the 3-inch-diameter pipe. Allow the bolts to self-tap themselves into the softer plastic for a tight fit, but do not overtighten them or you'll strip the PVC. Position and adjust the nuts as needed so there is a spark gap of approximately ¼ inch between the bolt ends inside the barrel.

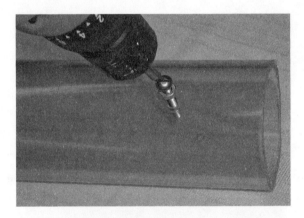

Step 2: Solvent-weld

1. Solvent-weld the 3-to-2-inch reducing fitting to the end of the 3-inch pipe closest to the electrode bolts. Clean the mating surfaces with PVC primer. Then attach the reducing fitting, carefully following the directions given on the PVC cement container.

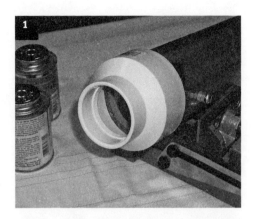

2. Solvent-weld the 3-inch threaded adapter coupling to the other end of the 3-inch pipe.

3. Solvent-weld 2-inch barrel to the corresponding opening in the 3-to-2-inch reducing fitting.

4. Let the assembly dry overnight.

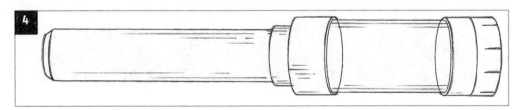

Step 3: Prepare the connector wires

Cut the 12-gauge wire into two 1-foot-long pieces. These will become the ignition wires. Crimp the spade connectors to one end of each ignition wire. Crimp the unmodified end of the barrel connector to the other end of the wire.

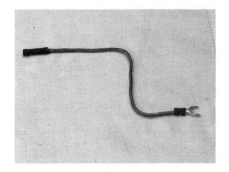

Step 4: Attach the stun gun

1. Attach the stun gun body to the gun barrel using the hose clamps, but be careful not to overtighten them. Position the stun gun body 90 degrees to the axis of the electrode bolts.

2. Attach the ignition wire spade connectors to the electrodes.

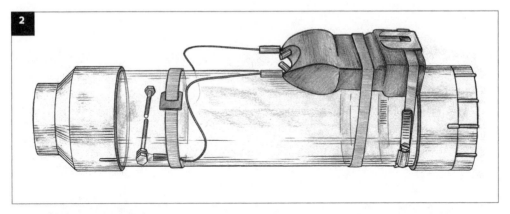

3. Attach the ignition wires to the stun gun electrodes using the modified barrel connectors. Crimp them on carefully.

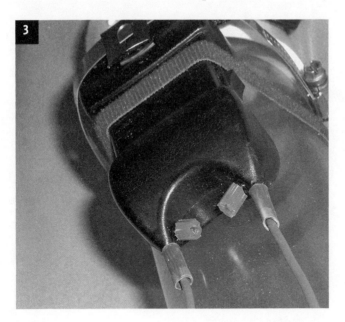

4. Cover the stun gun test leads (inboard electrodes) with wire nuts, which may need to be cut down in order to fit.

Step 5: Insulate

Insulate all exposed metal areas of the ignition path on the stun gun, and bolt the electrodes, with electrical tape or silicone sealant. It's easy for electricity to find its way underneath any insulation gap at the base of the electrodes. Cover the bolt connections with silicone sealant. You may want to wrap the stun gun body with bubble wrap, but the stun gun operates at such high voltage that the insulating wrap will not completely prevent shocks. Use caution to avoid contact between the stun gun's primary electrodes and your skin. Don't be the path of least resistance!

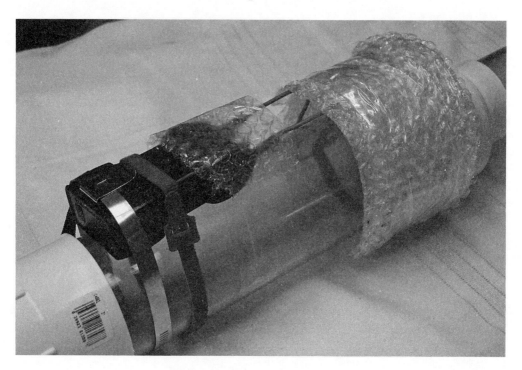

OPERATION

1. Remove the end cap, if it is screwed into the threaded adapter coupling.
2. Carefully push a potato into the cannon from the muzzle end. The sharp cutting edge ground into the muzzle will cut the potato into a plug of the correct size. The potato must fit snugly on all sides of the muzzle. Any gap will allow the expanding gas to "blow by" the potato. If that happens, the potato won't fly far.
3. Use the dowel or broom handle to push the potato plug 30 inches down into the cannon muzzle.
4. Direct a stream of aerosol deodorant or hair spray (check the label to make sure it contains hydrocarbons such as alcohol, propane, or butane) into the firing chamber (the 3-inch-diameter cylinder where the electrodes are). Start out with a 1-to-2-second burst. You'll likely need to determine the best amount of propellant by trial and error. Warning: The fluid stays in the combustion chamber and evaporates rather slowly. Always treat the gun as if there were flammable vapor in the firing chamber.
5. Immediately replace the end cap and screw it on securely.
6. Turn the stun gun on. Observe all safety rules.
7. Press the ignition button. Enjoy your work!
8. Many aerosol propellants contain chemicals that tend to gum up the inside of the cannon. Clean out the gun every few shots with a rag and nonvolatile cleaner. The end cap can get stuck from the gum. Use a wrench to unscrew it if necessary.

Swirl . . .

Ignite . . .

The pulse jet

The Jam Jar Jet

CHAPTER 4

The Pulse Jet

The first successful cruise missile, the Nazi-built V-1 (*V* stood for *Vergeltungswaffen*, which translates to "weapon of revenge"), was built by Germany during the middle years of World War II. The British called it the buzz bomb or the doodlebug because of the noise it made as it flew toward them. The engine that powered the V-1 wasn't a piston engine, and it wasn't a rocket motor. It was a unique type of jet engine called a pulse jet.

The pulse jet engine is simple, cheap, and powerful but isn't used in commercial aviation because it is incredibly noisy and vibrates like a gigantic, unbalanced chain saw. Pulse jets were invented in Europe in the early twentieth century but really had no practical use until the German scientist Paul Schmidt devised a dependable system of valves and fuel-delivery components for the Luftwaffe's terror weapon.

The V-1 flying bomb is an example of what's now called a cruise missile. A cruise missile is much like an automatically guided airplane: it has wings, a tail, and a motor. It is an "air-breather"; that is, it uses a fuel-air combustion cycle engine for power, not a chemical rocket engine.

Today the U.S. Air Force uses cruise missiles such as the Tomahawk, the ACM, and the Harpoon. Many countries build cruise missiles. For example, the French make the Exocet and the Chinese have the Silkworm. Cruise missiles are a popular armament because they are relatively easy and cheap to build.

The German V-1 was a tube about twenty-five feet long with two stubby wings extending from the fuselage. Carried piggyback on the rear was the pulse jet engine. The V-1 had a range of about 150 miles and could carry an explosive payload of nearly a ton.

THE V-1'S TEST PILOT

For the era in which it was constructed, the V-1 was a highly technical and difficult-to-engineer weapon. Much was promised, but time and again during its development the V-1 (officially called the Fi-103 by the Nazis) disappointed its designers by crashing during or just after takeoff. According to the engineering analysis, everything looked fine on paper. On the ground, all systems tested perfectly. But as soon as the Fi-103 roared down the ramp and took off, it lurched out of control, sending the engineers and their angry bosses diving for cover. The only way to figure out what was going on, it appeared, was to put a live test pilot onboard to observe.

Only a special person is willing to strap into a tiny cockpit on a jet-powered missile that has a propensity for crashing in a ball of flame. The Nazis would have to find a brave man indeed to attempt something so dangerous.

No man came forward. But Hanna Reitsch did. Reitsch was one of Germany's best, and bravest, test pilots. She was petite, young, brave, and fairly good looking (at least before the injuries she endured during numerous crashes). She test-piloted the Messerschmidt 163B, a dangerous and ultrafast rocket-powered airplane. She also flew the FW-61, the world's first helicopter, indoors at an auto show in Berlin in 1938.

An expert flyer, she was also, unfortunately, an unrepentant Nazi. She was a confidante of Himmler, Goebbels, and Hitler himself and was one of the last people to see Hitler alive in his Berlin bunker just before the Red Army stormed in.

Why such a bright woman would so wholly subscribe to such a hateful ideology is a mystery. More evidence of her perverse thinking is that toward the end of the war, she promoted the idea of airplane squadrons composed of "suicide pilots." The idea was to have fanatical German pilots fly modified V-1s into high-value targets such as Allied ships and headquarters buildings. Although the suicide squadron idea was never adopted by the German Luftwaffe, the concept was later embraced by the Japanese kamikaze corps.

At the German test site of Peenemünde, Flugkapitan (Captain of the Air) Reitsch strapped herself into the tiny payload section of a modified V-1 and flew four hazardous trips in the flying bomb. Three times she took off and could not find the answer to the V-1's problem. On the fourth try she determined that the bolts holding the wing to the V-1's body were breaking off when the pulse jet kicked in, causing the V-1 to become unstable and fly off course.

With the design problems fixed, the Germans cranked up production of the V-1 using forced labor from concentration camps. The Nazis aimed thousands of buzz bombs at London from launching areas in France, Holland, and other nearby countries. A sensor on the front of the V-1 determined roughly how far it had traveled since launch. When the preset distance was reached, fuel to the engine shut off and the V-1 soared down toward its target in a deadly, silent glide. The sound of the buzz bomb wasn't what scared people. It was hearing it stop that was frightening.

The first one hit London on June 12, 1943. At the height of their use, a few hundred were launched daily. But the British quickly became experts at intercepting the drones with Royal Air Force fighters and taking down many more with antiaircraft artillery. Consequently, only 25 percent of all V-1s launched ever reached their targets. The attacks ended when the Allies began to recapture land in occupied Europe, seizing launch sites as they went.

BUILDING YOUR OWN PULSE JET

You might believe that building a jet engine at home is beyond the means and capabilities of most hobbyists. But in this chapter, I'll show you how to build a simple jet engine—a pulse jet—for about fifteen dollars in materials and a couple of hours of labor.

Unlike, say, a turbojet or a fan-jet, both of which contain hundreds of precisely machined rotating parts, a person can build a working pulse jet in the garage with little more than hand tools and a welder. But even the need for those tools can dissuade some hobbyists. Those whose fabrication skills are limited to punching holes in sheet metal and bending wire can build a simplified version of the pulse jet: the Reynst combustor or, as it is better known, the Jam Jar Jet.

François Reynst was a Swiss engineer who experimented with a number of jet engine designs in the first half of the twentieth century. As a youngster he showed keen interest in the science of combustion. One day while experimenting with alcohol in glass jars, he accidentally discovered an interesting phenomenon. He found that if he placed a small amount of alcohol in a closed jar and punched a hole of a certain dimension in the lid, a weird pulsating combustion occurred when the fuel was ignited.

Reynst looked on with great interest as flames shot out of the hole. And

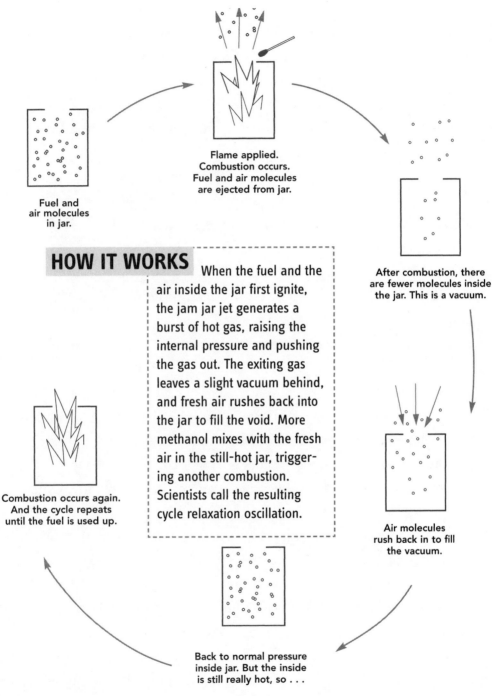

Fuel and
air molecules
in jar.

Flame applied.
Combustion occurs.
Fuel and air molecules
are ejected from jar.

After combustion, there
are fewer molecules inside
the jar. This is a vacuum.

HOW IT WORKS

When the fuel and the air inside the jar first ignite, the jam jar jet generates a burst of hot gas, raising the internal pressure and pushing the gas out. The exiting gas leaves a slight vacuum behind, and fresh air rushes back into the jar to fill the void. More methanol mixes with the fresh air in the still-hot jar, triggering another combustion. Scientists call the resulting cycle relaxation oscillation.

Combustion occurs again.
And the cycle repeats
until the fuel is used up.

Air molecules
rush back in to fill
the vacuum.

Back to normal pressure
inside jar. But the inside
is still really hot, so . . .

The pulse jet combustion cycle

then, almost magically, the flame was sucked back into the bottle and ejected again. The whole process would repeat until the fuel was expended. Reynst had a jar that was like the lungs of Saint George's dragon—it literally breathed fire. What young Reynst had discovered was a peculiar cyclical burner that would become known as the Reynst combustor. In actuality it was a simplified pulse jet engine.

Occurring within the jar is what scientists call systematic oscillation, which is caused by the burning of air and fuel and the geometry of the container. When the fuel and air inside the jar ignite, the Jam Jar Jet generates a burst of hot gas that raises the pressure inside the jar. The high pressure pushes the gas out of the hole in the jar in a strong pulse. But the exiting gas has mass and inertia, and when it leaves, too much of it exits. All of a sudden there's a vacuum inside the jar.

Nature, as the Dutch philosopher Spinoza said, abhors a vacuum. So fresh air rushes back in through the same hole to fill the void. The returning fresh air mixes with the residual alcohol still inside the jar. Since the inside is still very hot, it combusts again, and the hot gas is forced out again. The cycle repeats—explosion, thrust, vacuum—generating measurable amounts of thrust until the fuel is expended.

Reynst had invented a really radical type of pulse jet engine. Unlike any other, it has only one opening. But for a jet like this, that's all you really need. Continuous combustion cycles, such as those that that occur within a turbojet, require that the cycle's intake and exhaust be separate. But with pulsating combustion, the separation is not necessary, as the intake and exhaust parts of the cycle are not simultaneous. The combustion chamber "breathes" in and then out through the single port.

No two ways about it, this simple little contrivance, which can be constructed in less than a couple of hours on a kitchen table, is a type of pulse jet engine, and it operates on a real combustion cycle.

THE JAM JAR JET

The Jam Jar Jet consists of a pint-size mason jar, with a roughly conical, copper air-shaping ring suspended from the lid by wires. The lid has a round, carefully drilled half-inch-diameter hole in its center.

The jar itself is the see-through combustion chamber; the half-inch-diameter hole is the entrance and exhaust port. The conical copper ring serves a couple of purposes. First, its shape speeds up the exhaust stream and slows down the intake phase, providing better cycle efficiency. Second, the thermal mass of the copper absorbs heat, taking some of the thermal expansion strain off the glass walls of the jar.

SAFETY

This is a jet engine you are building—a tempest in a teapot, so to speak. I've never had any problems with this design, but because you're building it at home with parts you've procured yourself, no one can guarantee your safety. No one but you is responsible for your work. That said and understood, go forward with the project if you desire. If you do, here are the measures you should take to create a safe environment.

1. Do not experiment with different-sized jars and openings. Using too large a jar with too small an opening may result in an explosion rather than a combustion cycle.
2. Use neither more nor less fuel than directed.

(continued)

3. Use only the parts listed in the directions. People often ask me whether they can substitute something else for a part, or if they can make something a different size. My answer is always no, since I've not tried it or performed any analysis on that substitution.

4. Protect yourself by wearing safety glasses and gloves.

5. The Reynst combustor is an extremely efficient heating device. After just a few seconds of run time, the outside of the jar gets hot enough to burn skin. Do not touch the jar for five minutes after a successful run, and even then be sure to test it to make sure it is cool enough to handle.

6. The glass jar typically cracks after several successful runs. Once your jar cracks, carefully discard it by sweeping the entire assembly into a bag without touching it. Seal the bag and throw it in the trash. Jars are cheap enough, so just go get another one.

7. Keep spectators at a safe distance, at least eight feet away.

8. Always use a long-handled match or barbecue lighter to ignite the engine. After ignition, there is an immediate pulse of hot gas. A cigarette lighter or short match is not suitable.

9. Examine all parts before and after ignition for wear, and discard any worn parts.

10. Always use common sense before, during, and after running the Jam Jar Jet.

MAKING THE JAM JAR JET

TOOLS

Power drill

$1/2$-inch drill bit

File or sandpaper

$1/8$-inch drill bit

Wire cutter

Eyedropper or measuring spoon

Cookie sheet

Safety glasses

Gloves

MATERIALS

(1) Pint-sized mason jar with extra screw caps and lids

(4) 4-inch lengths of 22-, 24-, or 26-gauge magnet wire

(1) Conical air diffuser. You can make one out of copper pipe reducers. Purchase copper pipe fittings at a home center or hardware store. You'll need:

(1) $1^{1}/_{2}$-inch-to-$1^{1}/_{4}$-inch copper reducing fitting*

(1) $1^{1}/_{4}$-inch-to-1-inch copper reducing fitting

(1) Small bottle of methanol. The easiest place to buy a small bottle of fairly pure methanol is the local auto supply store. Many varieties of gas line antifreeze (e.g., brand names Heet and Pyroil) are mostly methanol. Check the label for ingredients before purchasing.

(1) Package of extra-long fireplace matches or a long-handled barbecue lighter

OPTIONAL ITEMS

Table salt

Boric acid crystals

Refrigerator

(1) 1-foot-long, 1-inch-diameter plastic or iron pipe

---

*These exact fittings may be hard to find. You may substitute combinations of other fittings to form a conical reducer of about the same size.

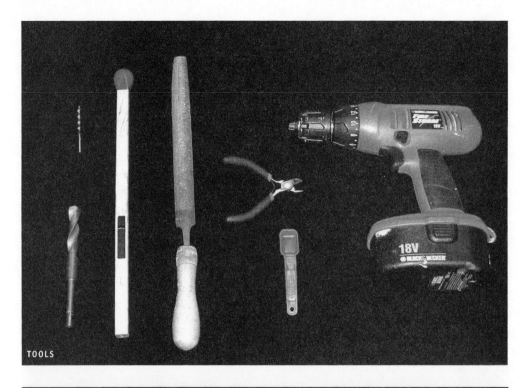

TOOLS

MATERIALS

CONSTRUCTION

Step 1

Drill a ½-inch-diameter hole in the mason jar's lid. Use a file or sandpaper to remove the burr. If the hole is jagged and cannot be made smooth and round, discard the lid and try again with a new lid.

Step 2

Drill four ⅛-inch-diameter holes in the smaller copper reducing fitting. The holes should be located about ¼ inch down from the edge of the small end of the adapter. Space the holes evenly around the perimeter at 90, 180, and 270 degrees from the first hole.

Step 3

Cut four 3-inch-long wires from thin-gauge magnet wire. Loop a wire through each one of the holes in the smaller copper reducing fitting, and tie them on. Extend the other end of the wires outward, radially, from the diffuser cone.

Step 4

Insert the large end of the small copper reducing fitting into the small end of the larger fitting. Press-fit them together firmly. This forms the conical air diffuser and heat sink. The total height of the diffuser should be less than half the height of the mason jar.

Step 5

Center the conical air diffuser in the middle of the jar. Crimp the wires over the edge of the jar so that the top of the copper cone remains suspended close to the top of the jar, with a gap of ¼ to ½ inch from the jar's lid.

Step 6

Sprinkle or use an eyedropper to insert 5 milliliters (slightly less than a teaspoon) of methanol into the bottom of the jar. At most, the methanol should just cover the bottom of the jar—you must use only a small amount.

Step 7

Screw the mason jar lid down onto the jar and the copper wires. The lid will hold the copper cone securely in place at the top of the jar.

Step 8 (optional)

Place the Jam Jar Jet in the freezer for two minutes. Based on my experience, slightly cooling the fuel and the jar seems to improve performance.

Step 9

Hold your thumb over the ½-inch opening in the lid. Swirl or shake the methanol inside the jar. Place the cooled Jam Jar Jet on a cookie sheet, and place the cookie sheet on a secure surface, away from flammable objects. When you remove your finger from the hole, you should notice a slight pressure release, and the jar will make a very faint *pffft* sound. If there is no slight pressure and no sound, swirl the fuel again. If there is still no pressure, there is a leak in the seal of the jar that must be fixed.

Step 10

Wear safety glasses and gloves. Using a long-handled barbecue lighter, hold the flame over the ½-inch opening.

The flame will ignite the fuel, and immediately the Jam Jar Jet will fire up. For the next five to fifteen seconds, the Jam Jar Jet will cycle, pulse, and buzz, operating on a low but audible frequency. With the lights down low, you'll enjoy a blue pulse of flame growing and shrinking under the lid as the jar breathes fire. The size and frequency of the jet pulses will vary according to conditions in the jar and in the atmosphere.

INTERESTING VARIATIONS

* Add a pinch of table salt to the methanol. This will cause the pulse to be bright yellow instead of intense blue.
* Add a pinch of boric acid crystals. The crystals will result in a deep green flame.
* Using pliers or a gloved hand, hold a tube ½ inch or so above the exit-entrance hole. Iron pipe or plastic works, and even a cardboard tube will last a little while. Experiment with the length and diameter of the tube. When the size is right, you'll be rewarded with a loud, deep, resonant buzz.
* Some Jam Jar Jet enthusiasts experiment with a metal jar instead of a mason jar, outfitting it with a resonant tube suspended above the hole.

This is sometimes termed a snorkler. Some snorklers have fuel-feed systems that drip methanol into the combustion chamber and thus can sustain combustion for long periods of time.

PROBLEMS?

If the methanol ignites with a single big whoosh instead of pulsing:
* Check the size of the hole, and make sure it is a full ½-inch diameter.
* Make sure the jar and methanol are not too warm. Put them in the freezer for a minute or two, and then try again.
* Make certain the jar is charged with the recommended amount and type of fuel.

If you hold the match over the opening, and it doesn't ignite:
* Methanol is hydrophilic and absorbs water readily from the air, so make sure the methanol is fresh. (The bottle's seal has not been removed for more than a day.)
* Cool down the jar in the freezer.
* Make sure you've got about a teaspoonful of fuel in the bottom, no more than that.
* Vigorously shake the fuel.
* Check the seal by listening for the *pffft* when you remove your finger from the hole. Rejigger the lid if necessary to get a good seal.

If the jar cracks
The Reynst combustor/pulse jet is a very efficient burner and therefore extracts a lot of heat from the fuel very quickly. If the jar you're using cannot handle the rapid expansion, it will crack. If this happens, carefully dispose of the broken jar and replace it with a jar with thicker glass if possible.

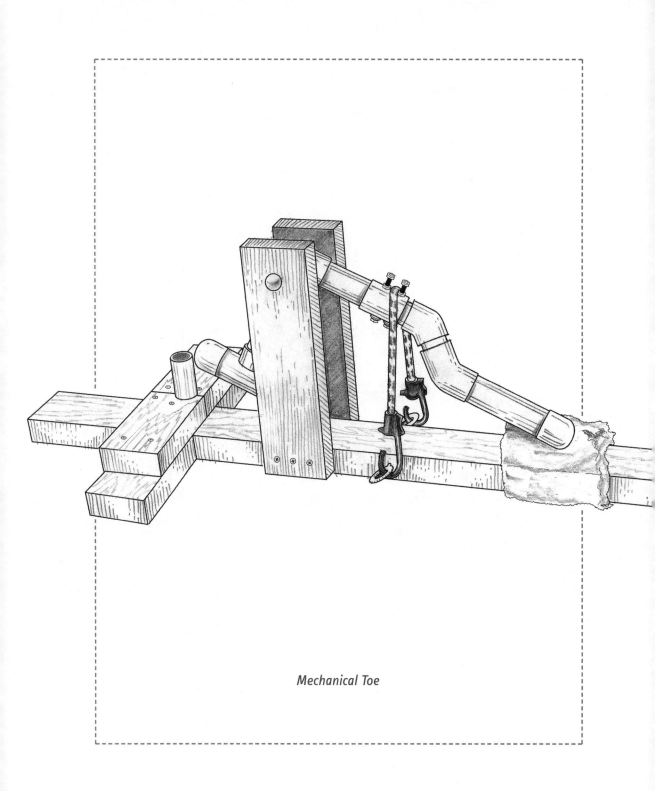

Mechanical Toe

CHAPTER 5

Hired Guns

Like the itinerant gunslingers of the Old West, there
have always been exceptionally clever scientists and engineers
willing to sell their expertise in weapon-making to the highest bidder. This
chapter features a couple of mechanically powered projects and also tells
the tales of a few of history's most famous "soldier-engineers of fortune."

In the mid-fifteenth century Europe was something of a mess. For the
previous hundred years the French and English had engaged in a costly
and bloody war that took thousands of lives and caused enormous suffer-
ing. The Black Death—bubonic plague—while perhaps not quite the
scourge it had been a hundred years earlier, was still killing thousands in its
frequent outbreaks. Perhaps nowhere in Europe, though, was life as cruel
as in Romania, where in 1456 Vlad III of Walachia ascended to power.
Vlad was a minor prince in the Balkan region of Europe. But the flinty-
eyed, black-mustachioed Vlad was a barbarian, a merciless and probably
mentally ill despot worthy of keeping warped company with the likes of
Caligula, Hitler, and Pol Pot.

British historian Robert Seton-Watson wrote that Vlad was "a man of diseased and abnormal tendencies, the victim of acute moral insanity." If contemporary reports are to be believed, he was a merciless despot who craved complete control and demanded complete subservience from his subjects. He was called Vlad the Impaler, Vlad Dracul, and Vlad the Devil, for he murdered thousands of people with little compunction and no mercy. His trademark act of barbarism was his preferred method of executing enemies, and the source of his nickname—impalement.

Vlad was a fierce fighter, and the soldiers whom his army captured usually regretted not fighting to the death. Vlad was said to enjoy eating dinner in his palace, surrounded by the bodies of a host of newly impaled prisoners. Besides captured soldiers, Vlad impaled priests, merchants, nobles, bureaucrats, Transylvanians, Hungarians, Walachians, men, women, children—pretty much anybody he didn't like, which was just about everybody. According to one account, he once saw a monk riding by on a donkey and, angered by some minor transgression now lost to history, impaled them both on a single stake.

By 1461, Vlad incurred the wrath of the powerful Ottoman sultan Mehmet II, known as Mehmet the Conquerer. Vlad had burned Ottoman villages and towns and impaled a reported 25,000 Ottoman civilians. Mehmet moved swiftly against the real-life model for Count Dracula (it was "Vlad Dracul" who inspired the vampire of Bram Stoker's novel), assembling a large and well-equipped army.

Setting out to find and exterminate Vlad, Mehmet assembled more than 100,000 men and two hundred ships, as well as his signature weapons, giant cannons, of which much more will be said shortly.

Vlad and his dark-hearted followers were hard to find and even harder to fight. They employed early guerrilla warfare tactics: frequent raids in which they materialized out of the woods, made a quick and deadly strike against the Ottoman Turks, and then disappeared. Vlad picked his targets carefully, often finding weak or poorly defended spots in the Turkish lines and inflicting great damage against Mehmet's much larger force.

Eventually the size and might of Mehmet's Turkish forces overpowered Vlad. The quantity of land troops, the ships, the cavalry, and the famed Turkish artillery cannonades drove the evil Vlad back to the remote mountains of Transylvania, where he continued to scourge his country.

After a couple years in Transylvania, the king of Hungary got his hands on Vlad and imprisoned him. Released from prison after just four years, he bided his time, hanging around royal courts for another twelve years under house arrest. His brother Radu "the Handsome" had become ruler of Transylvania. When Radu unexpectedly died, Vlad III made another play for power and invaded Walachia with a small army of supporters.

Mehmet remembered the destruction and ruin Vlad had unleashed just a decade earlier. Refusing to allow nearby Walachia to be ruled by the evil Vlad, Mehmet sent an army of well-equipped Turks, with artillery and specialized fighters, deep into Romania. The Ottomans did what they set out to do: Vlad was killed, perhaps in battle, perhaps by his own troops. In any event, his head was separated from the rest of his body, placed in a jar of honey, and sent to Istanbul as proof to Mehmet that Vlad Dracul—Vlad the Devil—was finally dead.

Mehmet the Conqueror in his prime

THE WALLS OF CONSTANTINOPLE

In most accounts of Turkish military history from Mehmet's time onward, cannons and other artillery play an important part. There is probably no battle so important to Ottoman history and no battle in which cannons played so dominant a role as the taking of the Byzantine capital of Constantinople.

In 1453 Constantinople was known as the Queen of Cities, the ancient capital of the Byzantines, and the place from which Emperor Constantine XI ruled a receding but still important empire. Despite the decline, the Byzantines continued to take pride in their huge and beautiful public works, monuments, and statuary. For centuries Constantinople was the largest city in the Western world. Unsurpassed in wealth and power, it was considered nearly impregnable. That reputation turned out to be its downfall. Its richness and apparent invincibility made it a target for the ambitious Mehmet II.

The Conqueror cared little about the theological value of the city's vaunted collection of religious icons and relics. But many were made of solid gold and silver and were festooned with huge gemstones and other valuables, making the city tempting for pecuniary reasons. Yet the city was surrounded by the legendary "Walls of Constantinople," a heavily fortified network of walls, earthworks, moats, and other barriers. Construction had begun in the fifth century, and for a thousand years the thick, fortified stone walls protected the city from attack. No city on earth was besieged more often (by the Huns, the Rus, the Turks, the Bulgars, and the Arabs to name but a few failed armies) than Constantinople and with less success. Those walls had proven themselves unbreachable time and again, an impenetrable bastion against usurpers and conquerors.

Walls or no walls, in 1453 Mehmet determined that Constantinople would be his. He gathered his huge army and went west to lay siege to the Queen of Cities.

Around the mid-fifteenth century siege warfare was being revolutionized by the invention of cannon. In the 1420s, after the Byzantines witnessed besiegers using crude cannons against them, they reduced the effectiveness of the new weapon by suspending bales of material, wood, and anything else that might absorb and diffuse a projectile's impact. But Constantinople's fifth-century fortifications were designed for spear-wielding infantry assaults supported by the occasional catapult barrage. They were no match for the new, modern weapons powered by the "Devil's Poudre"—gunpowder. Those once-invincible walls now presented a target of easy opportunity to the Turks.

Going into the siege, Byzantium knew that it would need modern technology to check its enemies. Considering his situation, Constantine decided to do something to close the gap in the race for medieval weapons of mass destruction. In 1451 he hired a Christian engineer—a Hungarian named Urban—to develop artillery for the Byzantines. But in a poorly considered decision, Constantine, rich as he was, refused to pay Urban adequately. Angered, the cannon expert sallied across the frontier to offer his technological skills to the Turkish side. Mehmet knew a bargain when he saw one and agreed to fully fund Urban's work. Thus, the long tradition of mighty Turkish artillery was born.

Urban's defection had a long-term effect. For hundreds of years—in fact, through the siege of Gallipoli in World War I—Turkish artillery would be among the most advanced and effective in the world.

Once Urban was in the employ of the sultan, he cast the largest cannon ever produced. The mammoth iron tube had a twenty-nine-foot-long barrel that fired stones said to weigh around twelve hundred pounds. The can-

The formidable Turkish cannon

non, as befit a device so large, was given its own name, or even two: it was sometimes called the Imperial and sometimes the Basilica. It was so heavy that a team of sixty oxen had to haul it to Constantinople from its foundry in Edirne, a city a hundred miles west of Constantinople.

Each time the cannon was fired, the expanding gases and friction wore and stressed its barrel to such an extent that it could be fired only seven times a day.

Nicolo Barbaro was an Italian physician in service to Constantine. An observant diarist and reporter, he kept careful notes about what was happening during the Ottoman siege. Urban's Imperial, he wrote, was very large: "The cannon was of exceptional size, throwing a ball twelve hundred pounds in weight, and when it fired the explosion made all the walls of the city shake, and all the ground inside, and even the ships in the harbor felt the vibrations of it.

"Because of the great noise, many women fainted with the shock which the firing of it gave them. No greater cannon than this one was ever seen in the whole pagan world, and it was this that broke down such a great deal of the city walls."

In addition to the cannon, the Turks employed many other techniques associated with medieval siege warfare. They dug tunnels under the walls and built tall siege towers that they rolled up to the walls.

The Byzantines mounted a vigorous defense. They dug countertunnels and threw burning material into those of the invaders. They poured hot pitch from the walls and set fire to anything wooden set against them. The smoke of fires, as well as the cannon, meant that the combatants fought without clearly seeing what was around them.

Mehmet had a force of 80,000 men at his disposal, regular troops and reserves, and 12,000 elite forces, perhaps the best fighters in all of Europe and the Middle East. These special fighters were called janissaries. Taken from Christian families as youths, janissaries were converted to Islam, then raised and trained in special schools to become fanatical fighters. They were the sultan's personal guard and most skilled soldiers.

Facing this army, Constantine could muster only about a quarter as many troops. Still, he did have those walls for protection. Perhaps, he thought, he could obtain help from European powers such as Venice, which might be sympathetic in the fight between a Christian empire and an Islamic one. But the Venetians offered only words of encouragement, no help of any real consequence. It became clear to Constantine that Byzantium would have to go it alone.

Sultan Mehmet stopped the cannon bombardment temporarily while an emissary was sent into the city to discuss surrender terms. If the emperor paid an annual tribute of 100,000 gold bezants, the emissary said, then Constantinople would remain Byzantine. Or all the citizens could simply take their possessions and leave, and no one would be harmed.

Unable to raise the money and unwilling to leave the city, Emperor Constantine and the Byzantines decided to tough it out. If Byzantium and Constantinople were to die, they would die fighting.

On May 29, 1453, Mehmet and his soldiers made a final, terrific assault at the location called the San Romano Gate. The battle began with the Turks "firing their cannon again and again, with so many other guns and arrows without number and shouting from these pagans, that the very air seemed to be split apart. . . . The janissaries," Barbaro continued, "and their officers attacked the walls of the poor city, not like Turks but like lions, with such shouting and sounding of castanets that it seemed a thing not of this world. The shouting was heard as far away as Anatolia, twelve miles away from their camp. . . . The people cried at the top of their voices, 'Mercy! Mercy! God send help from Heaven to this Empire of Constantine.' "

Bitterly, Constantine saw that no help was going to come. He had given up hope of assistance coming from the pope, from the Venetians, and now from God. The Turks fired their great cannon again and again. The janissaries poured through the breach in the walls, according to the diarist, "like hounds and came on behind the smoke of the cannon, raging and pressing on each other like wild beasts." Within fifteen minutes more than 30,000 Turkish troops were inside the walls, and the city was taken.

Constantine fought and died at the walls of his city. His head, according to several reports, was cut from his body and presented to Mehmet as a trophy. After eleven hundred years of power, the Queen of Cities fell to the guns of Urban and the troops of Mehmet the Conqueror. The end had come for Byzantium at the end of a long cannon barrel.

Long-barreled cannons, even those as large and powerful as Mehmet's Imperial, are simple and straightforward devices. Pared down to the basics, a cannon projects a slug or a shell toward a target via a long, straight aiming barrel. Military cannons use gunpowder, cordite, or other chemical propellants to impart energy to the projectile.

In the fifteenth century Sultan Mehmet and other powerful leaders were certainly taken with their armies' ability to produce the thunderous roar, fire, and smoke associated with gunpowder artillery. Their possession of powerful cannons, and the power that such devices bestowed upon their owners, explain their popularity.

Early gunpowder-powered cannons were expensive, labor intensive, and unwieldy to operate, but they played a key role in the armory of nearly every military power from the mid-fourteenth century onward.

But other types of cannons, ones that use no gunpowder at all, have been tested and used. In fact, there are numerous historical accounts of noncombustion-powered cannons. For example, the energy transferred from a rapidly unwinding spring, a discharging electronic capacitor, or an expanding quantity of steam could also shoot a projectile down a muzzle quite effectively. (In chapter 7 we'll explore the steam cannon in depth.)

In the following project, the energy stored in a flexible rubber membrane is transferred to a pellet to make a small but interesting toy cannon.

THE ELASTIC ZIP CANNON

The Elastic Zip Cannon is a cleverly designed toy hand cannon that uses stretched rubber fabric (a balloon) to power a small projectile, such as a dried bean, through a plastic pipe toward a target. It's a simple yet amusing project that consistently delivers good results.

SAFETY

The Elastic Zip Cannon is a modified slingshot powered by a stretched plastic membrane. The mechanism, while not particularly dangerous, does shoot peas and dried beans with enough velocity to sting. So protect eyes, and don't shoot it at people who don't want to be shot at. Also, follow the manufacturer's directions for working with PVC solvents and cleaners.

MAKING THE ELASTIC ZIP CANNON

TOOLS

Hacksaw

Small-diameter round (rat-tail) hand file

MATERIALS

(1) 12-inch-long, ½-inch-diameter PVC pipe

(1) 4-inch-long, ½-inch-diameter PVC pipe

(1) ¼-inch-long, ¾-inch-diameter PVC pipe

(1) ½-inch-diameter PVC tee fitting

(1) ½-inch-diameter PVC end cap

12-inch high-quality balloons

A handful of dried peas or beans.
  Larger dried beans, such as limas or
  garbanzos, work the best.

PVC primer and cement

TOOLS

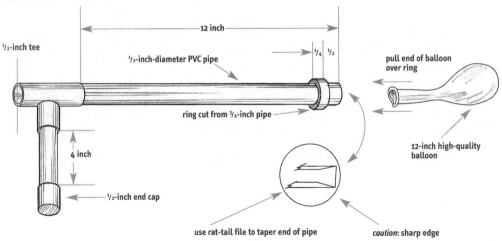

ZIP CANNON ASSEMBLY

12 inch

½-inch tee

½-inch-diameter PVC pipe

¼ ½

pull end of balloon over ring

ring cut from ¾-inch pipe

4 inch

12-inch high-quality balloon

½-inch end cap

use rat-tail file to taper end of pipe

caution: sharp edge

CONSTRUCTION

The Elastic Zip Cannon is quite straightforward and easy to make.

Step 1
Use the hacksaw to cut a 12-inch and a 4-inch length of ½-inch PVC pipe.

Step 2
Using the file, bevel the edges of the opening on one end of the 12-inch piece. A long, smooth transition works best. (See the diagram.)

Step 3
Take the ¼-inch length of ¾-inch-diameter PVC pipe, and slide it like a ring onto the 12-inch piece. You may slightly enlarge the diameter of the ¾-inch-diameter ring with sandpaper if necessary in order to slide it onto the 12-inch piece.

Step 4

Position the ¾-inch-diameter ring about ½ inch from the beveled end, and solvent-weld it into place, using PVC primer and cement.

Step 5

Attach the PVC tee fitting to the other end of the 12-inch-long piece (as shown in the diagram). Attach the 4-inch-long PVC handle to the tee. Then attach the end cap to the other end of the handle. Solvent-weld them into place.

Step 6

Slide the balloon over the ¾-inch diameter ring. Align the balloon such that its center line aligns with the center line of the 12-inch-long PVC barrel. The friction and tight fit will hold the balloon onto the barrel.

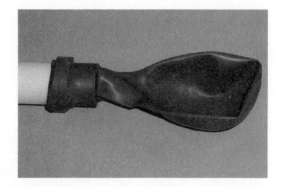

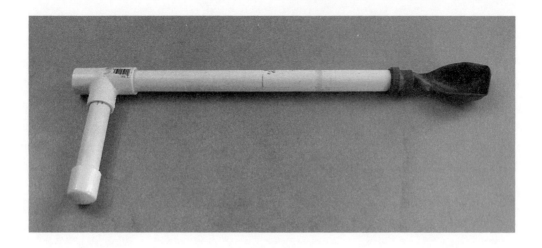

OPERATION

Hold the cannon vertically. Insert a bean into the barrel, and let it fall down into the balloon.

Level your shooter. Pinch the end of the balloon, and pull it back so it stretches. Aim the cannon at a target and release. The bean will shoot toward the target.

PROBLEMS?

If the bean gets stuck inside the barrel:

If the inside edge of the barrel, underneath the balloon, has any sharp edge at all, the bean will strike it and break apart or get stuck. Make sure the edge is smooth and long.

If the balloon breaks:

The balloon will not last forever and will break after repeated use. Balloons are cheap, so just replace them as required. But if your balloons break immediately, you may be pulling too hard, or you may be using poor-quality balloons.

⭐ SPORTS SCIENCE

Perfection is a rare thing, and examples of perfection in sports in particular are few and far between. But some actions are truly perfect because they are as good as they can possibly be. Gymnast Nadia Comaneci received the first perfect score awarded in an Olympic competition in 1976. In baseball, a perfect game means the pitcher handled every single batter he faced without allowing a hit, a walk, or an error. Literally, no one can do it better. Hundreds of thousands of major league baseball games have been played since 1880, and in that time there have been only seventeen perfect games. Very few bowlers, even those who bowl frequently, have rolled a 300 game. Those who have done so have attained perfection; no one will ever score better.

Football doesn't lend itself to perfection. There's always a chance that someone will score more points or do something better, faster, longer. Except for one record.

Back in 1969 the world champion New York Jets, led by Joe Namath, took on the Broncos at Mile High Stadium in Denver. The heavily favored Jets spent the first half of the game manhandling the Broncos. They built up a 15-point lead before the Broncos roared back and beat the Jets 21–19. The stunning loss was the talk of the sports world in the next morning's papers.

Almost lost in the big stories was the fact that little-known Jets punter Steve O'Neal set a record that still stands, one that may be as hard for future athletes to break as Joe DiMaggio's consecutive-game hitting streak, or Wilt Chamberlain's 100 points in a single basketball game. O'Neal, a Jets rookie, was sent in to punt the ball to the Broncos. With the ball on his own one-yard line, he was in a tough position. Normally an NFL punter stands fifteen yards behind the center. But with the ball on the one, he could stand back only eleven yards from the onrushing linemen.

The air is thin in Denver, and O'Neal, a good but in the final analysis not great punter, really got his foot into it. The ball traveled in an arching spiral trajectory for nearly seventy yards, sailing over punt returner Bill Thompson's head. It bounced near the Bronco thirty-yard line and finally rolled to a stop. On the one-yard line. O'Neal set a pro football record with a ninety-eight-yard punt that was four yards better than the previous record set in 1923 by Wilbur "Fat" Henry with the Canton Ohio Bulldogs, then in the National Football League.

If you think about it, ninety-eight yards is a perfect punt. It is as far as a ball can be punted. Since half-yards are not recorded in football records, a punt of ninety-nine yards or more from the one-yard line would result in a touch-back, netting only seventy-nine yards.

THE MECHANICAL TOE

The Mechanical Toe is a simple and oddly amusing device for propelling miniature footballs and tennis balls around the backyard. Unquestionably, you can throw a football farther than this contraption can toss it, but that's not really the point. The MT (Mechanical Toe) is easy to build, fun to use, and lends itself to a variety of experiments, some scientific and some simply amusing.

SAFETY

To kick a miniature football, the Mechanical Toe simulates the action of the human leg. The MT's power source is a stretched elastic cord, usually called a bungee cord. Bungee cords are capable of storing considerable energy when stretched. You must make certain that your cord is securely attached and in good condition prior to using the MT.

I suggest wearing safety glasses and gloves, at least during the initial test and the first few trials.

Clear the area in front of the MT, especially during test kicks, as the toe piece could come loose if not properly cemented.

Be sure to keep hands and feet clear from the impact area under the handle at all times.

MAKING THE MECHANICAL TOE

TOOLS

Handsaw

Hammer

Power drill

$^{11}/_{32}$-inch-diameter drill bit

Phillips head bit (if deck screws are used)

(1) $1^3/_8$-inch-diameter wood boring bit

Stapler or glue for cushion

Plastic spoon

MATERIALS

(3) 6-inch-long, 1-inch-diameter PVC pipes

(2) 2-inch-long, 2-inch-diameter PVC pipes

(1) 1-inch-diameter PVC tee fitting

(3) 1-inch-diameter PVC socket style (no pipe thread) end caps

(1) 1-lb. box plaster of Paris

(1) 36-inch-long wood 2 × 4 (the base)

(2) 12-inch-long wood 2 × 4's (the uprights)

(2) 6-inch-long wood 2 × 4's (the lower stabilizers)

(1) 12-inch-long wood 2 × 4 (the upper stabilizer)

(1) 1-lb. box $2^1/_2$-inch nails or deck screws

(2) size #2 eye-screws

(1) 2-inch by 2-inch thick foam or felt pad

(1) 90-degree 1-inch-diameter PVC elbow fitting

(2) 45-degree 1-inch-diameter PVC elbow fittings

(2) $2^1/_2$-inch-long #8 machine screws with 4 nuts

(2) 18-inch bungee cords, loops or hooks at each end

(1) 7-inch-long $^5/_{16}$-inch-diameter, fully threaded bolt, 3 washers, and 3 nuts

PVC primer and cement

(1) 3-inch-long, 1-inch-diameter PVC pipe

Assorted dowels and cardboard boxes for holding the football

TOOLS

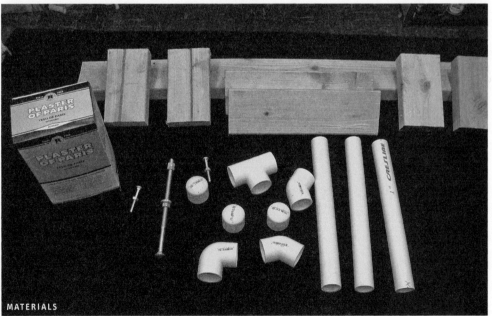

MATERIALS

Note, not all parts are shown.

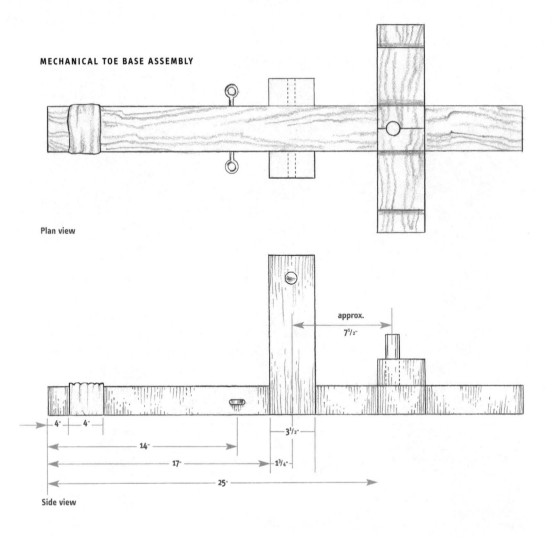

Plan view

approx.
7¹/₂"

4"   4"

14"

17"   1³/₄"

3¹/₂"

25"

Side view

CONSTRUCTION

Step 1: Cut PVC pieces to length

Measure and mark cutting lines for the three 6-inch PVC pipes and the two 2-inch PVC pipes. Use the saw to cleanly and squarely cut the pipe.

Step 2: Make the kicking toe

Solvent-weld one 2-inch-long pipe into each opposing end of the tee. (Leave the center hole of the tee open for now.) Solvent-weld the end caps onto each pipe end. Following the directions on the package, mix a sufficient quantity of plaster of 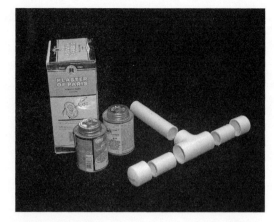 Paris to fill the tee and end cap assembly. Using a plastic spoon, fill the tee–end cap assembly with the plaster of Paris.

Before the plaster of Paris hardens, solvent-weld a 6-inch pipe into the middle hole of the tee. Put the toe aside, and allow it to dry and harden.

Step 3: Build the wooden base

Cut the 2 × 4's into lengths as described in the material list.

Drill two $^{11}/_{32}$-inch holes through the uprights. (See the wooden base assembly diagram.)

Attach the uprights to the base, using nails or deck screws. (See the wooden base assembly diagram.)

Carefully note the location of the upper stabilizer in relation to the base (as shown on the assembly diagram). Attach the upper stabilizers to the base using nails or deck screws.

Slide the lower stabilizers into position, and attach them to the upper stabilizer with nails or deck screws.

Attach the eye-screws to the base at the location specified in the assembly diagram.

Affix the cushion to the base at the location specified in the assembly diagram.

Step 4: Make the kicking leg assembly

Drill one clean $^{11}/_{32}$-inch hole for the pivot bolt through the 90-degree elbow fitting. (See the kicking leg assembly photo.)

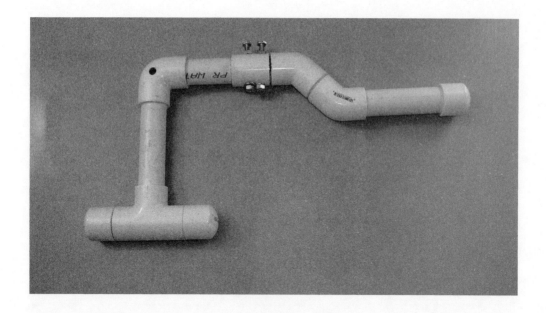

Solvent-weld the leg assembly together, as shown in the photo. Be certain to fully seat all connections and align all joints with the plane formed by the uprights and braces.

At the point shown in the photo, drill two $\frac{3}{16}$-inch-diameter holes through the leg assembly for the $2\frac{1}{2}$-inch-long, #8 bungee cord holder machine screws. The two holes are perpendicular to the $\frac{11}{32}$-inch pivot bolt hole.

Insert the two #8 machine screws through the holes, and bolt them into place. The space between the protruding machine screw heads is designed to keep the bungee positioned at the correct spot on the leg assembly.

Step 5: Mount the leg to the base

Align the holes in the base with the hole in the leg assembly. Insert the $\frac{5}{16}$-inch-diameter bolt. Place the washers, and tighten the nuts.

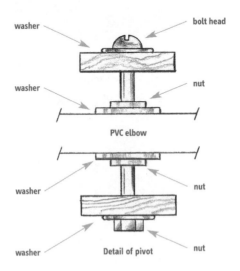

washer — bolt head

washer — nut

PVC elbow

washer — nut

washer — Detail of pivot — nut

Step 6: Make the kicking tee holder

Drill a $1\frac{3}{8}$-inch hole about $\frac{1}{2}$ inch deep in the middle of the upper stabilizer, at the point where the fully down toe will almost, but not quite, strike it.

NOTE: It may take a bit of trial and error to find the best point to drill the hole for the kicking tee. Place the assembled kicking leg in position on the wooden base prior to drilling the hole for the tee, and make certain the toe end will strike the projectile in the desired spot.

Place the 3-inch-long, 1-inch-diameter pipe in the hole. Wrap the bottom half-inch of the pipe with duct tape if necessary to obtain a good fit in

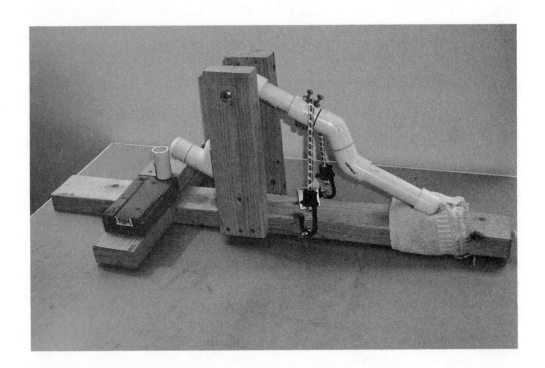

the hole. This is the kicking tee. The tee will holds the ball or other projectile. You will need to cut the kicking tee to length depending on the projectile. PVC tees of different lengths are needed for different projectile sizes. This arrangement will allow you to quickly and easily swap tees to accommodate different objects.

Step 7: Attach the bungee cord

Place the center of the bungee cord between the protruding screw heads on the leg assembly. Attach the hooks on the bungee cord to each eye-bolt on the base.

Step 8: Test the alignment

Check to be certain the handle hits the cushioned area on the base.

Check to be certain the bungee cord stays between the cord holder made from the #8 machine screws at all times.

Make sure the arm assembly pivots freely, but without excessive wobble or travel on the pivot bolt.

Make centering and tightening adjustments as necessary.

OPERATION

Place a handball, tennis ball, football, or other object on a kicking tee.

Lift the arm. As you do so, the bungee cord will stretch. Release the arm to let the toe strike and launch the projectile.

MODIFICATIONS

To increase the Mechanical Toe's performance, try several modifications:

1. Use a thicker bungee cord, or use two medium-thickness bungees.
2. Increase the weight of the toe.
3. Increase the radial velocity of the toe by lengthening the leg. Doing so will change the geometry of the toe. This will result in taller uprights and a longer bungee.
4. Use a harder ball.
5. Place the ball slightly off center to impart a spin to it. A spinning ball will travel farther.

ISAAC NEWTON AND FOOTBALL

Analyze any type of sporting activity, and sooner or later you will find yourself knocking up against some pretty basic physics. After all, a sport is often defined as "an activity that involves physical exertion and physical skill." Notice that you can hardly write *physical* without first writing *physics*.

The Mechanical Toe is a way to study the very human act of kicking a football using a mechanical model. The Toe lays bare the physics of kicking by doing away with the cumbersome human biology of skin, muscles, joints, foot bones, knee bones, and ankle bones. It allows you to focus explicitly on the physics.

Once a football leaves the foot or the hand of a player, two forces act on it: gravity and the drag caused by air rushing over its surface. In general, this rule holds true for all thrown, kicked, batted, clubbed, or otherwise flying balls. It is true in particular for football's basic motive modalities: the kick, punt, and pass.

What accounts for the different trajectories of kicks, punts, and passes? Many factors influence a football's flight, specifically, how far it travels, the time it spends in the air (in football parlance, its hang time), and how accurately it is directed to the spot where the punter desires it to go. Those factors are:

Launch speed
Launch angle
Air density and wind
Spin of the football
Shape of the football
Surface of the football

In the seventeenth century Isaac Newton first described the laws of nature that account for the behavior of objects in motion. These fundamental physics concepts are now known as Newton's Three Laws of Motion. These simple principles underlie all aspects of classical physics dealing with the motion of any object larger than an atom and going slower than near–light speed. Newton's Third Law of Motion is:

In every interaction, there is a pair of forces acting on the two interacting objects. The size of the force on the first object equals the size of the force on the second object. The direction of the force on the first object is opposite to the direction of the force on the second object.

The collision of the Mechanical Toe with the football is an interaction between two objects that have made intimate and robust contact with each other. So how does Newton's Third Law of Motion apply here? In the collision between the Mechanical Toe and the ball, both objects experience forces that are equal in magnitude and opposite in direction. The forces cause one object to speed up and the other object to slow down.

While the forces are equal in magnitude and opposite in direction, the acceleration of the objects is not equal. Because the toe and the ball have different masses, their collision results in unequal accelerations. The lighter object will have the larger acceleration.

If we could fill the Mechanical Toe with lead instead of plaster, then the more massive toe would impart even greater acceleration to the ball. If we could fill the toe with the same volume of an ultraheavy element such as thorium or uranium (plutonium would be even better!), then the toe would have even more mass, and the ball would really go flying!

Now let's analyze this kick by looking at Newton's other laws. Newton's first law says:

An object in motion will stay in motion and an object at rest will stay at rest unless acted on by a force. That is, once an object is in motion, it stays in motion until something happens to change it.

Newton defines a property of objects in motion called momentum, which is a consequence of its combined mass and velocity. The bigger and faster something is, the more momentum it has. So it is not incorrect to say that the game of football is mostly about using and controlling momentum, and coaches know this.

On most football teams, the punter is not an especially big guy—he'll be around 200 pounds, maybe 225. That may seem like a lot until you consider the players weighing 300 pounds who want to rush in and pound him to the ground before he can get his kick away. As soon as the center hikes the football, a stampede of defensive linemen make for the punter with a single aim: to slam him to the ground as quickly as possible.

Of course, there is an obstacle in the way of those defensive players. They have to push past the punter's cordon of beefy bodyguards—big, brawny guards, tackles, and ends who weigh an average 20 pounds more than even the defensive linemen.

To navigate through games with their incredible mass, skillful offensive linemen know a lot about leverage, physics, and momentum, although they may not think about these things in those terms. The players with the most mass have a weighty advantage. Their big muscles and even bigger gut help them keep the defensive players away from the punter. A really big football player can weigh more than 380 pounds, and he uses his mass like a physics pro to do his job, keeping his center of gravity low, his feet wide apart, and his shoulders in front of his body. Imagine attempting to push aside a moving 380-pounder in full pads. That's hard to do, because he's got a lot of momentum!

Newton's Second Law of Motion is just as important as the other two when it comes to football, for this is the law that describes the relationship between force and mass:

$$\text{Force (F)} = \text{mass (m)} \times \text{acceleration (a)}$$

Acceleration is the change in speed or direction as a result of force. The second law says bigger football players are harder to move than less massive football players. The Third Law of Motion says that the bigger the lineman, the more force is required to change that offensive lineman's speed or direction.

You may want to think of two grocery carts sitting in the supermarket aisle, one empty and one filled with food. The cart stuffed with groceries has more mass and will therefore be harder to move. Moving it takes more force. Likewise, making those giant football tackles change direction or speed takes more force. So a less massive player will have a tough time pushing aside a big offensive tackle to get his hands on the punter.

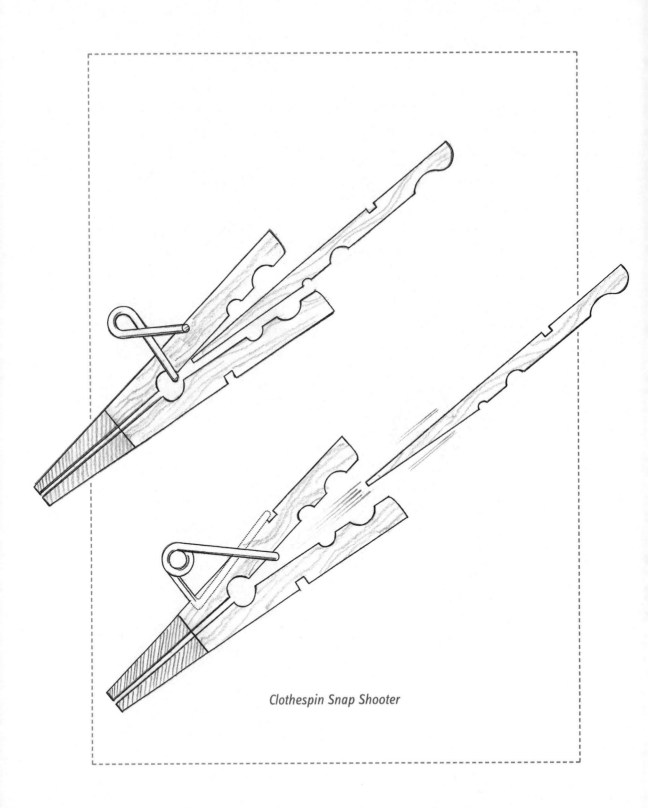

Clothespin Snap Shooter

CHAPTER 6

Mechanical Marvels

Suppose an experimenter has a small metal tube with a tight-fitting cover for each end. On a whim, she takes the metal tube and puts some fuel and oxidizer inside it. She tightly covers and fastens both ends and then ignites the fuel by applying heat or lighting a fuse. When enough energy is provided, or the burning fuse contacts the fuel-oxidizer mixture, the chemicals combine in a vigorous exothermic, or energy-producing, reaction. Rapidly expanding gas is created, and a great and perhaps uncontainable pressure builds up inside the tube. If the pressure exceeds the material strength of the tube, the whole thing blows up. This is, simply put, a pipe bomb.

Now instead of solidly fastening covers to both ends of the tube, suppose the daring experimenter simply slides a movable cover over one end. When she ignites the fuel, now what happens? Most likely the sliding cover flies off. It's a projectile, and she has made a primitive cannon.

The experimenter next leaves one end of the tube completely open and ignites the fuel while the tube lies upon the ground. When the chemicals ignite, the burning gas rushes out of the open end and propels the tube in the other direction. She has just built a rocket.

A cannon, a bomb, and a rocket are merely different manifestations of the same basic idea. The form that the device takes is simply a matter of details.

✴ THE AIR BULLET

In chapter 7, we will see how pressurized air can be used to project items through the air for considerable distances. For now let's consider another type of cannon, one in which air itself is the projectile.

It's hard to believe, but air (or smoke, or another vapor) itself can be made into a bullet of sorts. We're not imagining a simple stream of jetted air, but rather a real tangible projectile: an air bullet. In this chapter we will create a device called a Vortex Launcher that shoots a projectile made of air. The air bullet can be made visible, and you'll see that the bullet has a discrete size and shape and that it shoots through the air like an arrow (albeit a slow-moving arrow shaped like a doughnut).

The Vortex Launcher is capable of shooting a ring of air or smoke from one side of a room to another. It can blow out candles, knock down houses of cards, and set curtains aflutter. That's mere puffery compared to what the U.S. Army once conceptualized: an air bullet and vortex projector big enough to knock airplanes out of the sky.

During World War II the U.S. Army was interested in developing alternative methods of dispersing gases and vapors on the battlefield. The ugly truth is that the purpose of these experiments was to figure out "better" ways to use chemical weapons, although this technology might also have useful agricultural applications such as spreading insecticides. One method the army studied involved forming gases and vapors into high-speed traveling rings called vortexes. A vortex has the useful property of maintaining its shape—a doughnut—while it flies through the air.

Early in the war the army hired a well-known aircraft designer named Thomas Shelton to build a prototype vortex generator to prove the concept. Shelton was a highly regarded aeronautical engineer, well known for his futuristic wing and fuselage designs. In 1935 Shelton had become famous in aerodynamic engineering circles for designing the Crusader, a small, low-powered, teardrop-shaped airplane with two tails that was capable of cruising at 200 miles per hour.

For the U.S. Army project, Shelton built a bell-shaped chamber with a piston powered by a shotgun shell on one end and a hole in the other. He fired off the shotgun shell inside the chamber, and some pretty interesting things happened. An eighteen-inch-diameter ring of smoke shot through the air—traveling a full city block and making a weird howling sound as it whizzed by.

Shelton proposed building a larger and more powerful vortex launcher for use as an antiaircraft weapon. He suggested building one with a piston powered by a high-explosive charge. According to Shelton, such a vortex weapon could accurately fire a ring of air at supersonic speeds to an altitude of several thousand feet. The vortex would hit enemy aircraft and send them spinning, in theory causing them to lose control and knocking them out of the sky.

For a while the army funded some research into the idea. But ultimately it decided its research money should be spent elsewhere, and the air bullet capable of taking down an airplane was never developed. Nonetheless, the vortex gun did show up on a smaller scale in several toys, including the Flash Gordon Ray Gun (named toy of the year in 1949 by *Popular Mechanics*) and the Wham-O Air Blaster.

THE VORTEX LAUNCHER

This project is simple and inexpensive, but it shoots a pretty darn good vortex. And the bigger the vortex launcher, the bigger the vortex.

MAKING THE VORTEX LAUNCHER

TOOLS

Power drill with a 3-inch-diameter hole saw (Note: The exact size of the hole is not critical, but having a perfectly circular hole is important. A 3-inch hole saw is commonly available at hardware stores for cutting holes in doors for doorknobs. It may be possible to rent the saw.)

$3/_{16}$-inch-diameter drill

Large scissors

MATERIALS

12-inch-diameter round piece of thick plastic or rubber sheeting

Large plastic bucket, roughly 10 inches in diameter and 15 inches long (these dimensions are not critical)

3-inch long, $1/_4$-inch-diameter eye bolt with two nuts and two washers

Duct tape

TOOLS

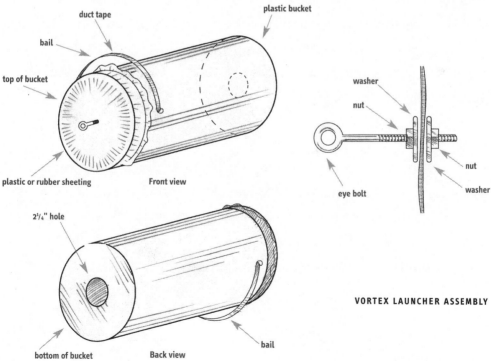

duct tape

plastic bucket

bail

top of bucket

plastic or rubber sheeting

Front view

washer

nut

eye bolt

nut

washer

2¼" hole

bottom of bucket

Back view

bail

VORTEX LAUNCHER ASSEMBLY

CONSTRUCTION

Step 1
From the piece of thick plastic or rubber sheeting, cut a round circular blank, of a diameter approximately 2 inches larger than the diameter of the open end of the plastic bucket.

Step 2
Drill a 3-inch hole in the center of the closed end of the bucket.

Step 3
At 2-inch intervals, use the scissors to cut radial lines into the plastic or rubber sheeting extending ¾ inch toward the center.

Step 4
Drill a ³⁄₁₆-inch hole in the center of the sheeting. Thread one of the nuts and one of the washers onto the eye bolt. Push the bolt through the opening, and place the other washer and nut on the bolt. (See the diagram on page 101 for placement.)

Step 5
Duct-tape the fabric to the opening, making certain that the perimeter is well sealed. The fabric should be applied fairly loosely so that there is adequate room for the membrane to travel. But too much travel will negatively affect performance. Trial and error is the best way to determine the optimum "tightness" of the fabric.

OPERATION

To operate the Vortex Launcher, pull back on the eye bolt, and quickly push it forward. How quickly should you operate the sheeting membrane? How far back should you pull, and how far forward should you push? The optimal technique is best learned by trial and error. Take note of the method that produces the tightest vortexes and best results.

INTERESTING OPTIONS

Target Practice

For practice, set up small houses of playing cards, and figure out what works best.

Visible Vortexes

The easiest way to see the vortexes as they shoot by is to fill the interior of the bucket with smoke. Blow smoke from a cigar into the bucket and then launch the vortex. I spray a fine mist of water into the interior along with the smoke. The moisture seems to make the rings hold together longer.

For nonsmokers, dry ice works well. Submerge a small hunk of dry ice in a container of water, and insert it into the cannon. This will produce copious visible condensation. Dry ice is dangerously cold and injures unprotected skin; always handle it cautiously and wear gloves.

Odoriferous Projectiles

Because the launcher has pretty good accuracy and launches fairly tight vortexes, you can direct a projectile of smell at unsuspecting individuals using the Vortex Launcher.

Place cotton that has been soaked with essential oils, such as peppermint or lavender oil (often available at drugstores, cosmetics stores, and natural foods co-ops). Let the smell concentrate for a minute inside the Vortex Launcher, then shoot.

✴ THE HELLBURNER OF ANTWERP

The lesson that Emperor Constantine XI learned the hard way (see Chapter 5)—that good engineers deserve good pay, especially if they are able to build machines that threaten your empire—is something that leaders should always remember in times of war. But the lessons of history are often forgotten. And evidently this same lesson was lost on the Spanish king Philip II.

Federico Gianibelli of Mantua was a military engineer of exceeding skill and cleverness. Around 1580 he was rudely rebuffed by Philip and the Spanish royal court when he offered them his engineering services. Angered by the patronizing attitude of the Spanish king, he went to Antwerp to work for the Dutch, who far more politely asked him to devise a defense to protect the city from an attack by the Spanish Armada. They backed up their request with cold, hard cash.

Gianibelli's device was the terrible-beyond-description "Hellburner." The Hellburner was a huge floating time bomb. A shell of a ship packed to the gills with explosives, incendiaries, and shrapnel, the Hellburner was designed to float into the middle of the besieging Spanish fleet with a time-delay fuse.

Historians describe Gianibelli as "purely a man of science and of great acquirements, who was looked upon by the ignorant populace alternately as a dreamer and a wizard. He was as indifferent to the cause of freedom as of despotism, but he had a great love for chemistry. He was also a profound mechanician, second to no man of his age in theoretic and practical engineering."*

Gianibelli's plan for defending Antwerp involved two ships, weighing approximately 75 tons apiece, which he requisitioned from the Dutch sailing fleet. The two ships were renamed *Fortune* and *Hope*, ironic names considering their destructive purpose. They were to become wave-borne volcanoes.

Gianibelli ordered workmen to place a solid twelve-inch-thick floor in the hold of each vessel. Built up the ship's sides from the floor was a bowl of masonry forty feet long and three feet high. The men working on the ship called this bowl "the crater."

The crater was filled with several thousand pounds of explosives, but not just any explosives. They used a new and secret type of gunpowder that had been compounded and ground by Gianibelli himself. Over the deadly crater workmen built a marble covering in the shape of a hollow pyramid. The pyramid was filled with stones, cannonballs, harpoons, iron hooks, random hunks of jagged metal, and every other available piece of junk that could be used as shrapnel. A wooden roof was constructed to camouflage the devilish payload.

Gianibelli's plan was fairly straightforward. The two Hellburners were to be steered down Antwerp harbor toward the massed Spanish Armada. At the right moment the explosives in the holds were to be detonated by an ingenious timing device, also designed by Gianibelli. The timer would cause a large spark from a flint and steel mechanism to ignite the mass of gunpowder in the ships' holds.

The weather on that momentous day in 1585 was mild. The sky was clear when the Dutch released the *Hope* and the *Fortune* from their moorings, sending them floating slowly on the tidal current down the harbor

---

*For more on this subject, see the excellent book written by John Lothrop Motley, a nineteenth-century British historian, called *History of the United Netherlands from the Death of William the Silent to the Twelve Years' Truce*.

toward the anchored Armada. Spanish naval officers saw the ships as they moved closer. Suspecting that they were in fact some type of floating bomb, they put their own ships and troops on alert. The *Fortune* sailed in first but ran aground near a bridge held by the Spanish. The Spanish watched uneasily as a small fire started on the deck of the ship. But it soon extinguished itself, and no further activity was visible.

The Spanish troops, peering out from behind defensive fortifications, started to laugh and jeer at the Dutch army's poor effort. But they also watched, perhaps a bit nervously, as the second ship, the *Hope*, floated toward them. The *Hope* lazily and innocuously rode the current toward the Spanish position, striking the bridge directly and coming to a stop.

Historians tell us that some of the Spanish sailors saw a thin curl of dark smoke rising from a fire burning on the *Hope*'s deck. Officers sent a team in small boats to investigate.

When the team boarded the ship, the mechanical clock controlling the fuse timed out, and a tremendous fireball erupted over the harbor. The *Hope* vaporized, as did the Spanish sailors on board, the bridge, and most of the Spanish troops, even those behind the fortifications. Nineteenth-century English historian and diplomat John Lothrop Motley describes the scene and its aftermath in detail:

> Houses were toppled down miles away, and not a living thing, even in remote places, could keep its feet. The air was filled with a rain of plough-shares, grave-stones, and marble balls, intermixed with the heads, limbs, and bodies, of what had been human beings. Slabs of granite, vomited by the flaming ship, were found afterwards at a league's distance, and buried deep in the earth. A thousand soldiers were destroyed in a second of time; many of them being torn to shreds, beyond even the semblance of humanity.
>
> The *Hope* had then destroyed herself, precisely at the right moment. All the effects, and more than all, that had been predicted by the Mantuan wizard

[Gianibelli] had come to pass. The famous bridge was cleft through and through, and a thousand picked men were blown out of existence. The Governor-General himself was lying stark and stiff upon the bridge which he said should be his triumphal monument or his tomb. His most distinguished officers were dead, and all the survivors were dumb and blind with astonishment at the unheard of convulsion.

The passage was open for the fleet, and the fleet [of Dutch allies], lay below with sails spread, and oars in the rowlocks, only waiting for the signal to bear up at once to the scene of action, to smite out of existence all that remained of the splendid structure, and to carry relief and triumph into Antwerp.

CLOTHESPIN SNAP SHOOTER

Gianibelli's contribution to military technology was a result less of innovation and invention than of his ability to make use of existing technologies in dramatic and powerful new ways. Fire ships had been used previously, but Gianibelli extended their range through clever innovations in everything from complex fusing to timers to the composition of the explosive.

Similarly, people have been making toy guns out of clothespins probably since spring-operated clothespins were first invented. Most often the pincerlike action of the clothespin is used as a trigger to release a stretched rubber band. The results are adequate but not terribly impressive. But like Gianibelli's Hellburners, the Clothespin Snap Shooter marks a definite upgrade in performance and sophistication over previous clothespin guns. The Snap Shooter makes several novel enhancements to a well-known device that turns it into a better-performing version.

In its extended form, this project has another parallel to the events of 1585. Like the Hellburner, the PG-17 version of the Snap Shooter incorporates the use of friction to set a wooden structure (a wooden match) aflame.

Just as Gianibelli figured out how to go from boat to bomb, this chapter will shows you how to go from clothespin to cannon. This project costs only pennies and takes less than a half-hour to build. Be forewarned, however, that it is almost addictively amusing. Upon seeing it work, most people will want to build their own immediately, so have a good supply of clothespins on hand.

The Snap Shooter has been around for a long time, perhaps not when the Spanish Armada ruled the waves, but certainly in the early 1970s. That's when Gordon Wells and Jerome Pohlen, from whom these plans originally came, built it as elementary school experimenters.

The Clothespin Snap Shooter is one of those rare, serendipitous occurrences that happen only when a number of unlikely things just happen to fall into place. It's like figuring out that bending a pickle fork in a certain manner turns it into a skeleton key, or that hooking the pickle up to just the right voltage source can turn it into a great nightlight.

It's amazing but true: the lowly spring-operated clothespin provides the perfect raw material for a gun that shoots small projectiles very well. A couple of quick part inversions, a small notch with a jackknife, and a long piece of tape are all it takes.

The Clothespin Snap Shooter is versatile. It shoots wooden matchsticks, Popsicle sticks, dried beans, or just about anything else that fits between the jaws of the gun and is shaped so that it can be flung out by the force of the rapidly uncoiling torsion spring.

1. The Clothespin Snap Shooter easily shoots across a fair-sized room. Don't aim it at people or breakable items.
2. The spring can pinch fingers when fired.
3. Remember that a burning match is dangerous. Use the PG-17 version outdoors and only in areas where a lit match will not start a fire. Use extreme caution when aiming. This device is not particularly accurate and may shoot in unexpected directions. Remove all flammable materials from the entire area before using.

MAKING THE CLOTHESPIN SNAP SHOOTER

TOOLS

Scissors

Utility knife

MATERIALS

(2) Wooden clothespins (the type with a coil spring that pinches two separate wooden levers together)

(1) 3-inch-by-5-inch cardboard index card

Transparent tape

Scissors

TOOLS AND MATERIALS

FIRING THE SNAP SHOOTER

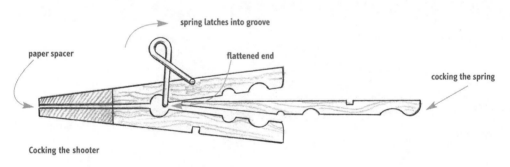

spring latches into groove

paper spacer

flattened end

cocking the spring

Cocking the shooter

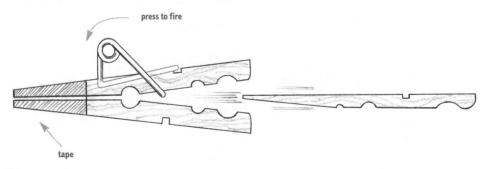

press to fire

tape

Firing

CONSTRUCTION

Step 1
Take apart the two clothespins, and lay the parts down on a table. There will be four wooden levers and two springs. Discard one of the springs.

Step 2
On one of the clothespins, rework the shape of the semicircular notch with a utility knife, until the notch has a flattened front end (as shown in the diagram). When the notch is shaped correctly, the end of the coil spring should hook just inside and stay without slipping.

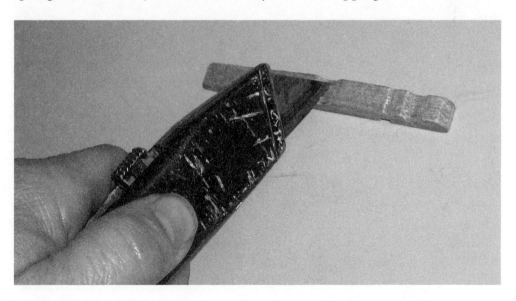

Step 3
With the scissors, cut two or three small rectangles of cardboard from the index card to form a spacer. Each rectangle should be the width of the clothespin, and the length of the distance from the thin end of the clothespin to the start of the semicircular notch.

Step 4

Slide the spring onto one of the wooden pieces, as shown in the diagram. One of the spring's bent ends must rest in the squared-off groove you just made in the wooden clothespin piece in Step 2.

Step 5

Arrange the two clothespin pieces and the cardboard as shown in the diagram. Tape the ends into place using a few turns of strong packaging tape.

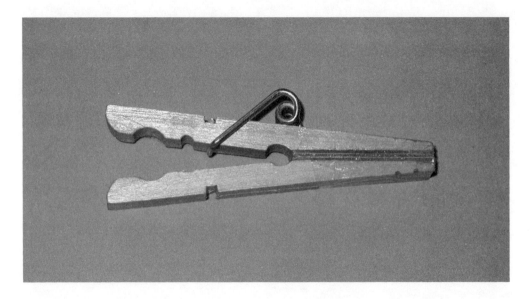

Step 6

That's it! The Clothespin Snap Shooter is ready for action.

FIRING THE SNAP SHOOTER

Step 1

Cock the spring by taking the remaining wooden clothespin piece and pushing the bent metal end of the spring toward the semicircular notches. The coil spring will rise up, away from the top of the clothespin. When the bent end of the spring reaches the notch, it will fall into the notch and latch into place. (The diagram shows how this is done.) The device is armed and ready to fire.

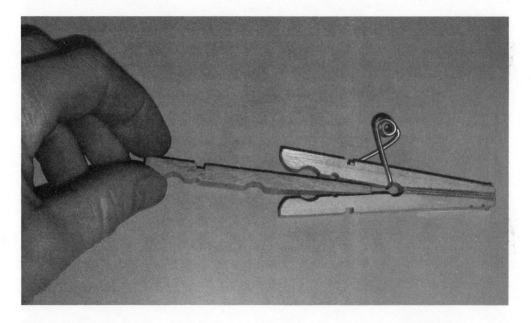

Step 2

Troubleshooting tip: If the bent spring end will not stay in the semicircular groove, add more layers of cardboard between the two flat clothespin pieces. Doing so will change the geometry of the notch. You may need to experiment with the cardboard and the angle of the notch until the geometry is just right.

Step 3

Place the Snap Shooter on a flat surface, spring side up. Insert the narrow end of the pushing stick (it is also the projectile) into the jaws of the shooter until it just contacts the cocked spring end.

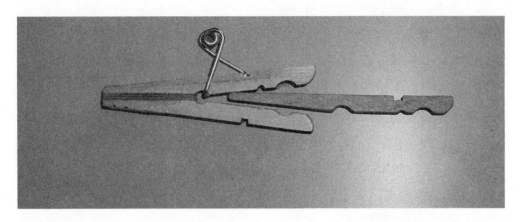

Step 4

Trigger the shooter by gently pulling on the spring coil. When the spring leaves the notch, it pushes against the projectile, firing it across the room.

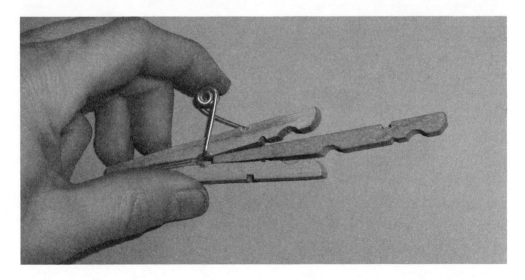

THE PG-17 VERSION

ADDITIONAL MATERIALS

Strike-anywhere matches (fresh, new matches work best)

CONSTRUCTION

After the shooter is cocked, put a strike-anywhere match inside the jaws of the Snap Shooter. When fired, the match will often ignite and fly through the air ablaze.

ADVISORY: When the match head is correctly placed against the jaws of the shooter, the motion of the jaws will ignite the match. Use out of doors only. Only adults or children with adult supervision should attempt this. Use only in a safe area!

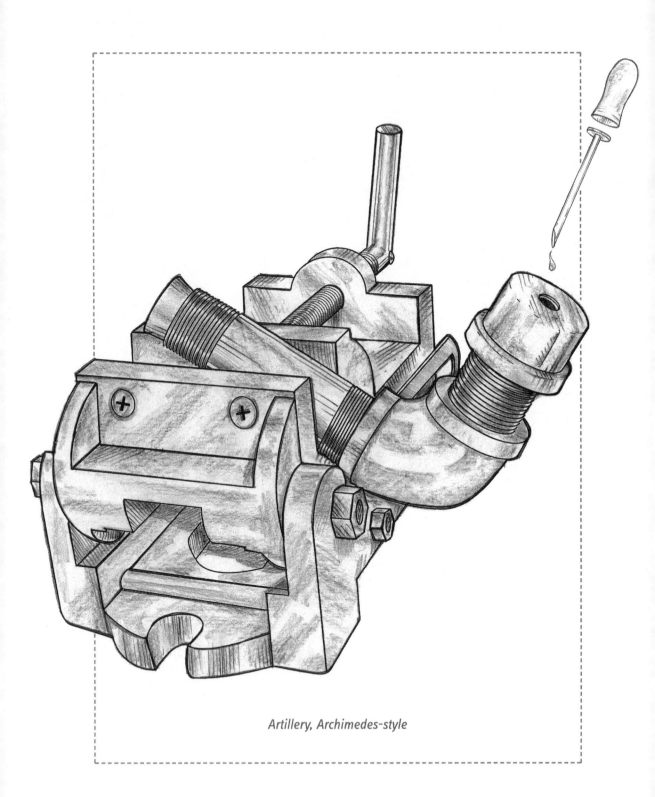

Artillery, Archimedes-style

CHAPTER 7

Extraordinary Ordnance

It is not always easy to determine the true provenance of a world-changing invention. In fact, the history of invention and discovery is frequently clouded by controversy as to which person or group deserves credit. For instance, Alexander Graham Bell is widely believed to be the inventor of the telephone, but a case can be made that it was developed earlier by others. There are quite a few claimants, including Elisha Gray, Johann Philipp Reis, and Antonio Meucci.*

Another inventive controversy concerns the television. Two individuals claim paternity: Vladimir Zworykin, a Russian-born American inventor working for Westinghouse, and Philo Farnsworth, a privately backed inventor from Utah. Another controversy, well known among historians,

---

*In June 2002 the U.S. House of Representatives passed a resolution to honor the life and achievements of the nineteenth-century Italian-American inventor Antonio Meucci. The bill states, "We have all grown up believing that Alexander Graham Bell invented the telephone. However, history must be rewritten if justice is to be done to recognize Meucci as the true inventor of the telephone."

involves the invention of radio. While most history teachers give credit to Guglielmo Marconi, a very strong case based on patent information can be made for Nikola Tesla, who claimed that he was truly the father of radio.

These disputes are familiar to those who study the history of science. When it comes to television, radio, and telephones, establishing the rights to such products involves huge amounts of money. But controversy can arise even for the invention of things whose only value is bragging rights. Case in point: the Wave.

The Wave is a well-known occurrence at sporting events. An exercise in crowd psychology, it involves a wave of jumping humanity sweeping around a stadium or sports arena. The Wave starts when a group of people leap to their feet with their arms up and then immediately sit down. Under the right circumstances, people next to the initiators do the same thing, as do those next to them, and so on. Viewed from above, a synchronized wavelike movement rolls through the audience, momentarily disturbing the equilibrium like a gust of wind across a field of Dakota wheat.

Two different groups claim to have invented the Wave, and both attempt to back up their claim.

While one frequent claim is that the Wave was first performed in Mexico City during the 1986 World Cup, it was almost certainly in use at sporting events in the United States prior to that.

George Henderson (aka Krazy George), a professional cheerleader, claims he invented the Wave at an Oakland Athletics baseball game in October 1981. He states on his website (www.krazygeorge.com), "I, Krazy George, invented, created and otherwise started the most famous fan participation routine in the history of the world."

Opposing Krazy George is a group of University of Washington alumni who say their school's cheerleaders orchestrated the first Wave during the third quarter of a Washington-Stanford football game, also in October 1981.

Even if determining with certainty the one true date and place where it first began is impossible, the Wave is interesting in ways beyond its ability to exhort teams to play better or to keep fans interested in a lopsided game. It reveals interesting relationships between human psychology and the physics of waves.

Professor Tamás Vicsek of Eötvös Loránd University in Budapest, Hungary, analyzed videotapes of fourteen waves in large football stadiums. He found that human waves can be modeled mathematically and conform to rather rigid physical rules of behavior, even though they are composed of the short-lived actions of thousands of people with no personal connection to one another, save the fact they root for a particular sports team. Vicsek found that Waves usually propagate in a clockwise direction, move at a speed of about 27 miles per hour, and are about 25 feet wide, which corresponds to an average width of fifteen seats.*

New research pertains to the critical mass of Wave initiation: What does it take to start a wave? Specifically, how many people does it take to start one? The answer lies in a property that Vicsek terms the "activation threshold."

The activation threshold is a measure of how likely the average person in a crowd is to stand up and do a Wave if the people around him or her stand up. The activation threshold is affected by a host of factors including the density of the crowd (i.e., the number of empty seats) and the level of excitement within the stadium.

Vicsek found that when conditions are right, a Wave can be initiated by fewer than two dozen people standing up simultaneously. In less favorable conditions, an initial group of forty people may be required to start the phenomenon.

---

*Students of the stadium wave phenomenon may ask, Why do stadium waves go clockwise? Does this have something to do with the Coriolus effect? Do stadium waves travel counterclockwise in the southern hemisphere? No, they almost always go clockwise because most people are right-handed.

THE T-SHIRT CANNON

In addition to the Wave, other nonfootball-related phenomena occur at events that involve cheerleaders. Many observers feel that few things excite crowds at sporting events, more than getting something, anything, for free. At the pinnacle of this measure is the moment when a cheerleader marches onto the field with perhaps the most coveted of all human possessions: a free T-shirt.

At one time the delivery of valuable booty was limited to the distance a human arm could throw it. This limitation provided spectators in rows one through fifteen near the playing field with an inestimable advantage. But with modern technology, even a fan in the nosebleed section has a chance to grab a hurled T-shirt. The enabling technology? The T-shirt cannon.

Although the word *cannon* sounds dangerous, no statistical records detail people being injured by a ballistic T-shirt. That's not to say, however, that no one has been hurt in stampedes to grab them.*

Presented here are plans for designing and fabricating a T-Shirt Cannon. May your aim be true and your trigger finger quick.

---

*Maude Flanders, a fictional character on television's *The Simpsons*, was fatally injured in a February 2000 episode in a freak T-shirt cannon incident.

SAFETY

Many of the same safety and legal concerns associated with the Night Lighter 36 (see Chapter 3) hold true for the T-Shirt Cannon. For example, T-shirt Cannons may not be legal in your area. So again, you should first check with local authorities to see if any regulations pertain to the construction and use of T-Shirt Cannons.

Second, devices that use pressurized air can be dangerous. They can shoot with great power, and in the worst-case scenario they can fail dangerously.

1. The T-Shirt Cannon is designed to shoot T-shirts. Shooting other any objects, especially near people, is not recommended.

2. Do not fire directly at people. If the cannon is used to distribute T-shirts to crowds, use a high firing angle when aiming the gun, so the shirt falls down gently on people. Make sure people are aware of your activities before you use the gun. Hitting an unprepared person, even with something as soft as a T-shirt, can cause injury.

3. Because of the high air pressure inside the tank, always use care and wear eye protection.

4. Don't operate a damaged or worn T-Shirt Cannon. Check the cannon frequently for signs of wear.

5. Don't take any construction shortcuts, use inferior materials, or make substitutions in the materials list.

6. PVC's mechanical properties become impaired in extreme weather. Don't use a T-Shirt Cannon in temperatures below 55 or above 95 degrees Fahrenheit.

(continued)

7. The vapors from PVC cement are flammable. Allow all joints to dry fully before using.

8. In general, despite best efforts, things occasionally go wrong. If you follow the directions here carefully, the chances of a bad outcome are remote, but unexpected things can happen: material flaws, bad judgment, pure bad luck. Always use common sense, and take careful note: *Build and use this device at your own risk.*

Working with PVC Pipe

Rules and techniques for working with PVC pipe were provided in Chapter 2. Review those instructions before undertaking this project.

It is especially important to review the general safety instructions for PVC solvent, which are on page 36. Finally, read and follow all chemical instruction labels carefully.

MAKING THE T-SHIRT CANNON

TOOLS

Tape measure

Hacksaw (for cutting PVC pipe to size)

Power drill

Drill bit (sized to fit tank valve)

Screwdriver

Pliers

Utility knife

Wire stripper

Soldering iron (recommended)

TOOLS

MATERIALS

The Barrel Assembly

(1) 25-inch-long, 3-inch-diameter schedule 40 PVC pipe

(1) 32-inch-long, 2-inch-diameter schedule 40 PVC pipe

(2) 1-inch-diameter "close" cast-iron pipe nipples*

(1) 3-inch-to-1-inch PVC reducing coupling, with female pipe threads on the 1-inch end†

(1) 2-inch-to-1-inch PVC reducing coupling, with female pipe threads on the 1-inch end‡

(1) 3-inch white PVC coupling, one side smooth, one side threaded

PVC primer and cement

Plumber's Teflon tape

(1) 1-inch-diameter U-shaped water sprinkler solenoid valve (Orbit Irrigation Model 57224 or equivalent, available at Home Depot and other places)

(1) 3-inch-diameter white PVC threaded end cap

(1) Air tank valve, $\frac{1}{8}$-inch, NPT, minimum of 1 inch long. (For example, look for a Schrader Bridgeport "Camel" Model 38-900 or a Tru-Flate 47-951. These are typically available at large hardware stores. Alternatively, go to www.mcmaster.com and search for "$\frac{1}{8}$-inch tank valve."

(1) O-ring. (The O-ring seals the valve body to the end cap tightly against air leakage. Therefore, the diameter should be just large enough to slide onto the tank valve body. O-rings are available at most larger hardware stores.)

(1) 3-foot-long, 1-inch-diameter wooden dowel (pushing stick)

The Electrical Assembly

(3) 9-volt battery holders (available at Radio Shack)

(1) 2-inch-by-2-inch-by-5-inch project box (available at Radio Shack)

(3) 9-volt batteries

(3) 9-volt battery wire connectors

(1) toggle switch with on-off label plate (Radio Shack 275-602)

(1) momentary switch (Radio Shack 275-609)

Low-voltage insulated wire

Pressure gauge (like a bicycle tire gauge)

Tie wraps

----

*A "close" pipe nipple is the shortest type of nipple and is threaded from each end with the thread meeting in the middle.

†Note: A 3-inch-to-1-inch reducing coupling is hard to find at hardware stores. You may substitute combinations of other available fittings, as long as you are ultimately able to go from a 3-inch-diameter PVC pipe to the 1-inch-diameter pipe nipple.

‡These reducing fittings can also be hard to find at hardware stores. You may substitute other available fittings, as long as you ultimately are able to go from a 2-inch-diameter PVC pipe to the 1-inch-diameter pipe nipple.

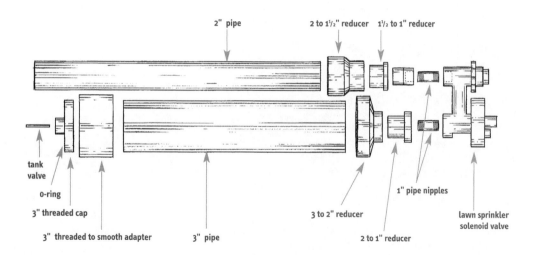

2" pipe

2 to 1½" reducer 1½ to 1" reducer

tank
valve

o-ring

3" threaded cap

3" threaded to smooth adapter   3" pipe

3 to 2" reducer

1" pipe nipples

2 to 1" reducer

lawn sprinkler
solenoid valve

BARREL ASSEMBLY

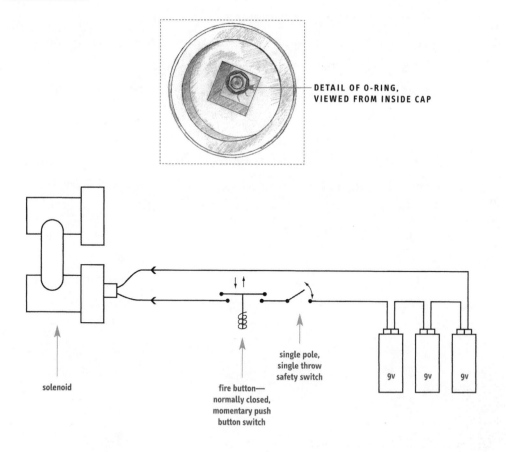

**DETAIL OF O-RING,
VIEWED FROM INSIDE CAP**

solenoid

fire button—
normally closed,
momentary push
button switch

single pole,
single throw
safety switch

9v   9v   9v

ELECTRICAL ASSEMBLY

CONSTRUCTION

Step 1: Cut the PVC barrel pieces

Measure and mark a cutting line 25 inches from one end of the 3-inch-diameter PVC pipe. Use the hacksaw to cleanly and squarely cut the pipe.

Now, measure and mark a cutting line 32 inches from one end of the 2-inch-diameter PVC pipe.

Step 2: Assemble the barrels

Assemble a combination of reducing fittings that connects the 3-inch-diameter PVC pipe to the 1-inch-diameter pipe nipple.

Similarly, assemble a combination of fittings for the 2-inch-diameter PVC pipe to the 1-inch-diameter pipe thread connection.

Solvent-weld the smooth end of the 3-inch threaded coupling to the 3-inch-diameter PVC pipe.

Step 3: Attach the barrels to the solenoid valve

Wrap the pipe threads with plumber's Teflon tape. Screw the barrel assemblies into the openings on the valve. There will be just enough room to mount the 2-inch and 3-inch barrels to the solenoid valve simultaneously.*

----

*Some brands of solenoid valve may not be large enough to handle both barrels at the same time. There must be at least 3$\frac{1}{4}$ inches of space from centerline to centerline in the solenoid valve to accommodate the barrels.

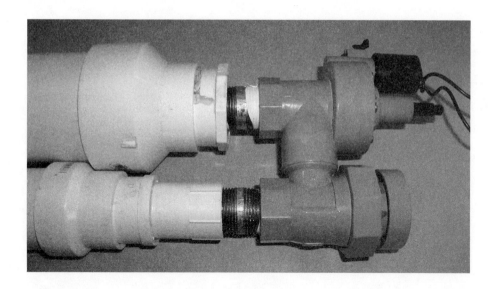

Step 4: Attach the air tank valve

Drill a hole in the center of the threaded end cap just large enough to enable the valve body to be screwed through the PVC of the end cap. Once it's in place, slide the O-ring onto the valve body, and tighten the nut. Tighten the nut down on the O-ring enough for a good seal, but do not overtighten.

 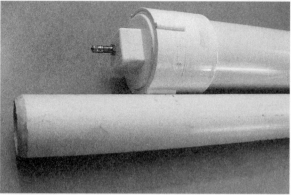

Screw the threaded end cap into the threaded end of the 3-inch-diameter coupling, using the Teflon tape to seal the connection tightly.

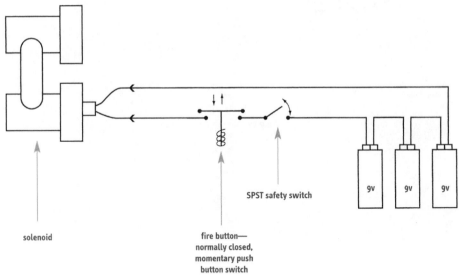

SPST safety switch

9v 9v 9v

solenoid

fire button—
normally closed,
momentary push
button switch

ELECTRICAL ASSEMBLY

Step 5: Construct the battery and switch box

Mount the battery holders in the project box. Place the batteries in the battery holders, and attach the 9-volt wire connectors to the batteries. Wire the batteries together in series to make a 27-volt power source. This will provide enough voltage to operate the 24-volt solenoid on the sprinkler valve.

In the top of the project box, drill a hole just large enough for the toggle switch (the safety switch) and another hole for the momentary-on push button switch (the trigger switch). Insert the switches through the holes, and secure them using the nuts supplied with the switches.

Wire the 27-volt battery supply, the toggle switch, and the trigger switch together in series. (See the diagram.) Attach the solenoid leads to each end of the circuit.

Test the circuit by turning the toggle switch to "on" and depressing the push button. If wired correctly, the solenoid valve in the sprinkler valve will make a "thunk" sound when the trigger is pushed.

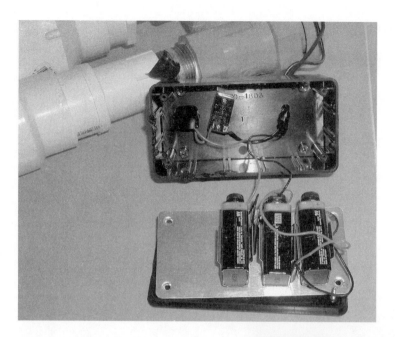

Step 6: Make final connections and attachments

When the circuit works correctly, solder all connections and close the project box. Attach the project box to the gun barrel using the tie wraps, as shown in the diagram.

Allow all of the PVC connections to cure overnight.

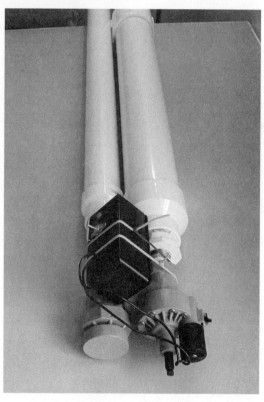

Step 7: Pressure-test the gun

Using a bicycle pump or other pressurized air source, apply approximately 50 psi to the device. Check all of the connections and attachments by brushing a soapy water solution on all of the joints. Bubbles indicate the presence of an air leak. Eliminate any leaks by tightening the pipe joints or reseating the O-ring. Properly constructed PVC joints should not leak. Leaks occurring at solvent welded pipe joints are uncommon but may be difficult to fix. Try to remedy leaking joints by applying more solvent.

When the gun holds pressure and the solenoid valve operates when the trigger is pushed, the T-shirt Cannon is ready for use.

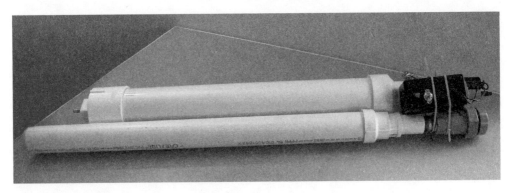

T-shirt cannon ready for use

OPERATION

Step 1: Load the T-shirt

With the safety toggle switch in the "off" position, load the cannon. Place the T-shirt in a plastic bag, or tightly wrap and fold the T-shirt so that it fits snugly but freely into the mouth of the barrel, and then tie it with tape. If the T-shirt unrolls during flight, it will not travel very far.

Push the T-shirt down into the barrel with the pushing stick.

Step 2: Pressurize the T-shirt Cannon

Using a hand pump or compressor, pressurize the air chamber of the cannon to 70 psi. Check the pressure inside the tank with a pressure gauge. Do not exceed 70 psi. Higher pressures may be unsafe.

Step 3: Fire the cannon

Move the safety toggle switch to the "on" or "fire" position. Carefully aim the cannon, and press the red trigger button.

PROBLEMS?

If the T-shirt Cannon doesn't fire, or if it fires weakly, try the following fixes:

* Make sure the pressure in the air tank is 70 psi.
* Check for leaks using soapy water solution.
* If you cannot accumulate pressure in the air tank, check the position of the manual release switch on the solenoid valve to make sure it is closed. Make sure all pipe connections are tight and the O-ring is well sealed.
* Make sure the solenoid valve is wired to the trigger button correctly. Remember, you should be able to hear the solenoid operate when the "fire" button is pushed.
* The solenoid valve must open rapidly. Check the operation of the solenoid.

THE ARCHITRONITO

The Architronito, Leonardo da Vinci's steam cannon, was first described in detail in one of the many notebooks da Vinci jotted down his ideas in. Apparently, he was inspired by the Greek scientist Archimedes, who first devised a steam cannon to protect his home city of Syracuse from invaders. Archimedes' weapon used steam power to hurl a hard projectile with nearly as much velocity and momentum as a gunpowder-powered cannon.

The Architronito model described here is simple, consisting of a length of iron pipe, a few fittings, a water injector, and a heat source. The really interesting thing here is the incredible amount of energy contained in the volume of steam produced from a very small measure of water. When water flashes to steam inside the steam cannon, enough energy is produced to easily shoot a rubber stopper across the room. Imagine the power available from a full-sized steam cannon!

Leonardo da Vinci, although most well known as a painter, actually made his living as a military engineer. During his lifetime (1452–1519), a number of city-states on the Italian peninsula, including Naples, Siena, Milan, and Florence, warred with one another. Because of the turbulent political environment, clever engineers were in high demand. Da Vinci designed on paper a number of highly original, novel weapons, but most were never used by combatants. In Leonardo's notebooks are many drawings and sketches including designs for the first flintlock rifle, a 135-foot-wide rock-throwing ballista, and a rapid-firing repeater crossbow. Of these ideas, perhaps none is as interesting as the one for a steam-powered cannon.

Why was Leonardo interested in developing such a device? At the time, the exact composition of gunpowder was a closely guarded secret, and much of the powder was of poor quality. The availability of powder was also a problem. As we have seen, gunpowder is a mix of sulfur, charcoal, and potassium nitrate (also called saltpeter). In gunpowder the sulfur and charcoal burn with oxygen—not from the air, but from the chemical bonds within the saltpeter.

In Renaissance Italy saltpeter was a rare and therefore expensive commodity. In the medieval period its only known sources were dung heaps. Only in the nineteenth century were alternative sources—guano (also dung, but from sea birds) from South America and salt lakes in Chile—found. Today we make saltpeter from synthetic nitric acid, so no one has to go scrabbling around in sewage.

Because of the difficulty and cost involved in making gunpowder, Leonardo sought an alternative. His steam cannon design placed the cannon's breach (the closed end) in a pot of hot coals. An iron ball was placed in the muzzle of the cannon. Then a small amount of water was introduced, through a system of injectors and vents, into a small but extremely hot gap behind the ball. The water would flash to steam, expanding thousands of times in an instant. With a great hiss and whoosh, the steam would hurl the cannonball in the direction of the enemy.

Leonardo wrote about the quantities of water required, the size of the projectiles, and how far they could be fired. As it turns out, only a small quantity of water was required to hurl the projectile. Steam is a very energetic source of motive power for cannonballs. Based on the data in his notebook, it appears that da Vinci performed some experiments on the cannon. Most likely, though, the weapon was never put into battle, as there are no historical records describing its use.

Preceding Leonardo's steam cannon machinations was the precursive device designed by the brilliant and eccentric geometer Archimedes. A catapult that used steam power, it was, in principle, a cannon. Leonardo learned of it from a manuscript written by the Roman author and statesman Cicero. Italian scholar Francesco Petrarch (1304–74) found the manuscript in a church library. Petrarch collected Greek and Roman manuscripts that had been neglected for many centuries in various libraries.

The Cicero manuscript proved to be of great interest to Leonardo, who used the information to sketch the device he called the Architronito in honor of Archimedes. Apparently he produced only drawings of the steam gun. In 1981 Ioannis Sakas, a Greek expert on the work of Archimedes, used this information to build an accurate replica consisting of a steam chamber, valve, and muzzle.

Sakas built a hot fire under the steam chamber. When the temperature inside reached 400 degrees Celsius, Sakas introduced a very small amount of water, which he found produced enough rapidly expanding steam to throw a tennis-ball-sized stone fifty yards.

The steam cannon concept was not widely used in actual warfare until the British Royal Navy placed steam-powered grenade hurlers on board some of its ships during World War II. The Holman Projector was a 4-foot-6-inch-long pipe on a swivel mount connected to the ship's steam boiler through a pedal-operated valve. The general idea was that, once low-flying enemy aircraft began an attack, sailors would drop a live hand grenade down the pipe. They'd tug and manhandle two large iron handles to aim the swiveling pipe at the attacking Messerschmidt. At the instant the airplane swooped in, they'd stomp on the pedal. A tremendous blast of high pressure steam would gush through the valve body, propelling the hand grenade toward the enemy aircraft and, they hoped, knocking it out of the sky.

It was a clever idea—simple, easy to use, and cheap. But, most sailors who had occasion to fire the Holman Projector had a poor opinion of it. It was cumbersome to use and hard to aim. Many sailors thought the Holman was more dangerous to themselves than to the enemy. But there are reports of sailors shooting potatoes at other British ships as a form of recreation, possibly making the Holman the earliest-known embodiment of potato cannon technology.

You can build your own simple steam cannon, an Architronito model made from pipe and some kitchen utensils.

SAFETY

Despite its relative simplicity, this project still requires great care to perform safely. It necessitates working with very high temperatures, delivered by way of a propane-powered heating torch. When done correctly, a small rubber cork will be shot from the end of the Architronito like a very small cannon-ball. There could be a small spray of steam and hot water as well. Plan accordingly, and clear the area in front of the cannon. This is good advice for all cannons.

Take responsibility for making sure that no one is in the line of fire who could be hit either by the stopper or by hot water. Also remember that the metal pipe becomes very hot. Bystanders must be warned against and prevented from touching the hot pipe.

1. It takes a bit of experimentation to determine the optimum pressure with which to press the cork into the muzzle. Start with a light but firm push, and vary the pressure later based on earlier results.

2. Use only the tools and materials listed in the directions.

3. Protect yourself by wearing safety glasses and gloves. Both are very important.

4. Keep spectators at a safe distance, at least eight feet away, and make sure the Architronito is not aimed at them.

5. Keep a minimum area of fifteen feet in front of the Architronito clear of people and anything that could be damaged.

6. Follow all manufacturers' directions regarding use of the heating torch and fuel.

7. As always, demonstrate common sense before, during, and after the project.

MAKING THE ARCHITRONITO

TOOLS

Propane-powered heating torch

Power drill and bit the same diameter as the injector tip

Vise

MATERIALS

This project is composed from plumbing fittings available in most hardware stores. If you need help finding or identifying nipples, caps, and other pipe fittings, ask for help at the store.

(1) ³/₄-inch-diameter black iron pipe cap

(1) ³/₄-inch-diameter black iron pipe nipple, approximately 3¹/₂ inches long overall

(1) ³/₄-inch-diameter 90-degree elbow

(1) ³/₄-inch-diameter close black iron pipe nipple

(1) Rubber stopper, size 00 (10mm diameter at bottom, 15mm at top)

Cup of water

Rubber bulb from a small glass eyedropper

Hollow steel injector tip from a turkey baster or flavor injector

TOOLS

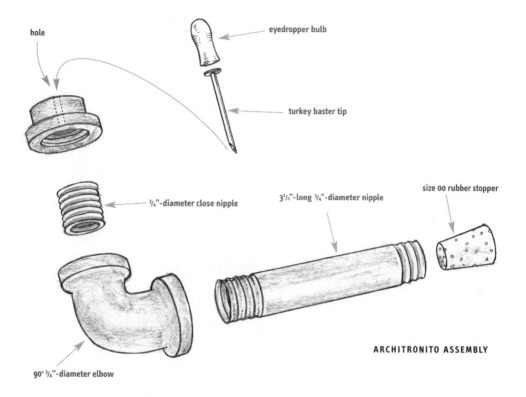

hole

eyedropper bulb

turkey baster tip

3/4"-diameter close nipple

3 1/2"-long 3/4"-diameter nipple

size 00 rubber stopper

90° 3/4"-diameter elbow

ARCHITRONITO ASSEMBLY

CONSTRUCTION

Step 1

Drill a hole in the center of the pipe cap the same diameter as the hollow steel injector tip.

Step 2

Assemble the steam cannon pipe fittings as shown in the assembly diagram. Screw all the connections down securely so there are no leaks.

Step 3
Assemble the water injector by affixing the rubber cap from the eyedropper to the steel injection needle from the turkey baster or flavor injector.

OPERATION

Step 1
Put on your gloves and safety glasses. In a steel vise, clamp the assembled steam cannon, as shown in the diagram.

Step 2
Heat the elbow with the flame of the propane torch for about a minute or until the elbow is very hot.

Step 3
Draw water into the flavor injector by squeezing the bulb. Only a small amount of water is necessary.

Step 4
Place the rubber stopper into the open end of the $3\frac{1}{2}$-inch pipe nipple, and press it into place.

Step 5
Insert the injector tip firmly in the hole in the end cap. Fire the cannon by squeezing the bulb to inject water into the cannon.

Step 6
Remember, the cannon gets very hot. Let the cannon cool undisturbed until it is safe to handle.

TIPS

1. Make certain the tip is inserted securely in the hole and there is no space for pressure to leak around the needle. If a leak is present, rotate the injector. If that doesn't help, discard the end cap, and make a new one using a slightly smaller-diameter drill bit.

2. You can increase the distance the stopper travels by using increased pressure to place the stopper in the nipple. But too much pressure may result in a misfire. If a misfire occurs, let it cool, then carefully remove the cap with tongs or gloved hands.

3. If the cannon does not fire, check to make certain the elbow has been adequately heated, that enough water is injected, and that the steel tip of the injector is firmly seated and no pressure leaks exist.

There are myriad ways to shoot projectiles beside using nitrogen-laden chemical propellants. We've explored projectile shooters powered by springs (the Clothespin Snap Shooter), compressed air (the T-Shirt Cannon), and steam (the Architronito). The following project uses the most personal power source of all, air from human lungs.

★ THE MARSHMALLOW SHOOTER

The Marshmallow Shooter has probably has been around since mini-marshmallows were invented. This is a favorite of schoolchildren everywhere. Simple to build and fun to use, this project will appeal to children of any age.

SAFETY

There aren't too many safety issues with this project. Wear safety glasses. Don't shoot people in the face with the marshmallows. Don't eat dirty marshmallows. That's about it.

MAKING THE MARSHMALLOW SHOOTER

TOOLS

No tools are necessary for this project!

MATERIALS

(1) 10-inch-long, $\frac{1}{2}$-inch-diameter PVC pipe

(6) $4\frac{1}{2}$-inch-long, $\frac{1}{2}$-inch-diameter PVC pipes

(2) $\frac{1}{2}$-inch-diameter PVC end caps

(2) $\frac{1}{2}$-inch-diameter PVC tee joints

(2) $\frac{1}{2}$-inch-diameter PVC elbow joints

(1) $\frac{1}{2}$-inch adapter, one side threaded, one side smooth

(1) $\frac{3}{4}$-inch-to-$\frac{1}{2}$-inch coupling, the $\frac{1}{2}$-inch side threaded, the other smooth (for the mouthpiece)

Bag of mini-marshmallows

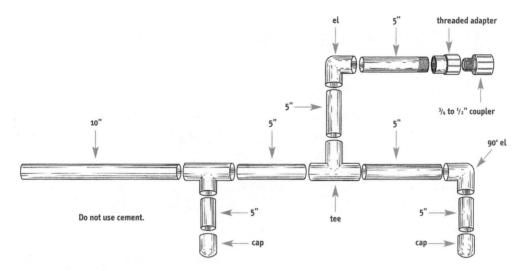

el   5"   threaded adapter

5"

³/₄ to ¹/₂" coupler

10"   5"   5"   5"   90° el

Do not use cement.   5"

tee   cap

cap   5"

MARSHMALLOW SHOOTER ASSEMBLY

CONSTRUCTION

Fit all the parts together as shown in the diagram. Note that no PVC cement is required or desired. Simply dry-fit the pieces together.

OPERATION

1. Remove the mouthpiece from the gun.
2. Load a mini-marshmallow (or two) into the barrel. Replace the mouthpiece.
3. Bring the gun up so that you can blow into the mouthpiece. The mouthpiece swivels so you can adjust it to suit your preference.
4. Blow a sharp burst of air into the mouthpiece. The bigger and shorter the duration of the huff, the farther the marshmallow will go.

THE PHYSICS OF BLOWGUNS

The blowgun converts human energy into projectile motion. Inside the barrel of the blowgun, an expanding push of breath pushes against the fins of the dart, causing it to accelerate down the tube.

A question that people sometimes ask is, How long should the blowgun be in order to obtain the greatest velocity? That's a good question, and one we'll attempt to answer intuitively. At what point does the dart reach its greatest speed?

Some people believe the dart is traveling fastest in the barrel just after the mouthpiece. They argue that the blast of air starts and stops suddenly, like a trumpet player playing a sixteenth note. That line of reasoning posits that after the initial push, the fins rub against the pipe walls, causing friction and slowing the dart until it clears the barrel.

In actuality, the amount of air huffed into the pipe is very large in comparison to the total volume of the tube. The volume of a 6-foot-long, $\frac{1}{2}$-inch-diameter blowgun is

$$\text{volume of blowgun} = \pi \, (.25^{\wedge}2) \times 72 = 14 \text{ cubic inches}$$

In anatomy studies scientists have found that the average volume of air in a medium blowgun huff is more than 60 cubic inches. For jungle blowgun hunters the huff is likely much greater. In fact, the huff from a typical hunter's lungs is so large in comparison to the tube volume that the dart accelerates continually down the entire length of the pipe, reaching maximum velocity as it leaves the muzzle.

So would a 10-foot blowgun be a better weapon? Would a 15-foot gun? The short answer is yes. The long answer is maybe. A hunter would impart greater velocity and accuracy to his dart with a longer blowgun. But he or she must figure out a way to conveniently carry it while hunting and must figure out a way to build such a tube out of a material strong and stiff enough to keep it from drooping and wobbling.

✸ THE YAGUA BLOWGUN

Blowguns have been used for hunting small mammals, birds, lizards, and even fish for thousands of years. These devices are known throughout the world, although the cultures of the Amazon basin and the Malay archipelago are most strongly associated with them.

In the hot, steamy rain forests, native hunters learned to use whatever nature provides in the way of tools, food, and weapons. And they found blowguns well suited to their needs. Perhaps because of the heavy vegetation and relative smallness of prey animals, the highly accurate but low-power weapon became a favorite method of hunting.

Blowguns go by a variety of names. In England, where they are used purely for sport, they are known as *blowpipes*. In South America the gun is called the *pucuña*. The jungle residents of Southeast Asia refer to them as *sumpitan*. Japanese ninja warriors called them *fukiya*, and Cherokee Indians gave them the name *tugawesti*. In all these places blowgun hunters took this activity very seriously, practicing until they could hit a small, fast-moving target at incredible distances. One report says that some Borneo tribes were so serious about blowgun hunting that they removed the upper incisors from young boys' mouths to improve their blowing ability.

Although we won't go as far as the Borneo tribesmen and require dental work, the next project does have a wild, primitive flavor. Carefully note the cautions associated with the construction and use of the blowgun. Remember, a blowgun is not a toy. It is a hunting tool of considerable power, a demonstration of physics, and a display of cultural anthropology.

MAKING THE BLOWGUN

TOOLS

Hand saw

Utility knife

MATERIALS

(1) 4-foot-long, $\frac{1}{2}$-inch-diameter PVC pipe

(1) $\frac{3}{4}$-inch female pipe thread to $\frac{1}{2}$-inch straight socket PVC reducing fitting

(1) 1-inch straight to $\frac{3}{4}$-inch male pipe thread PVC reducing fitting

PVC primer and cement

Tape or fabric for grip (optional)

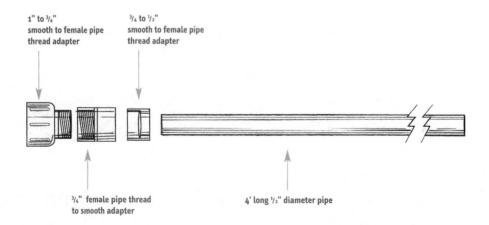

1" to $\frac{3}{4}$" smooth to female pipe thread adapter

$\frac{3}{4}$ to $\frac{1}{2}$" smooth to female pipe thread adapter

$\frac{3}{4}$" female pipe thread to smooth adapter

4' long $\frac{1}{2}$" diameter pipe

BLOWGUN ASSEMBLY

CONSTRUCTION

A blowgun is simple to build. Blowguns in South America and the Pacific are usually fitted with a mouthpiece, which can be a simple widening of the breach end, a separate funnel-shaped wooden cup, or a piece of hollow bone. The purpose of the mouthpiece is to concentrate the entire force of the blow behind the dart.

1. Cut the PVC pipe to a 4-foot length.
2. Smooth both ends of the cut pipe with a utility knife, removing burrs.
3. Using PVC cement, attach the ½-inch socket end of the reducing fitting to one end of the pipe. Do not get cement on the threaded areas.
4. Screw the 1-inch straight to ¾-inch male pipe thread reducing fitting onto the ¾-inch female threaded fitting.
5. Optional: Wrap a band of grip tape or fabric 12 to 18 inches from the mouthpiece (depending on your preference).

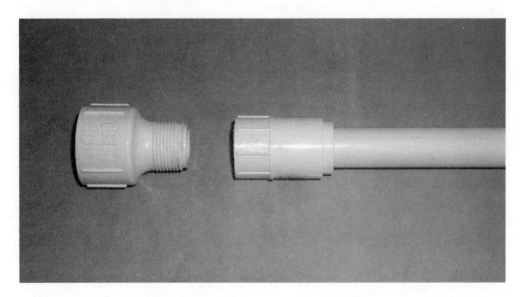

WIRE DARTS

Stun darts are relatively heavy, have blunt ends, and don't pierce skin or animal hide upon impact.

Stun darts are easily made out of a wire nut and wooden match. A wire nut is a plastic electrical connector commonly sold at hardware stores. First choose a wire nut that fits the inside diameter of the barrel as closely as possible. Then insert a wooden match into the nut socket and twist it into place. The matchstick will add stability during flight.

Wire darts have thin, hard shafts and are used in small-game hunting.

MAKING THE WIRE DARTS

MATERIALS

(1) 5-inch-long wire cut from coat hanger, or a cotton swab with the cotton ball removed from one end. Paper swabs work adequately, but better performance is obtained using a longer, wooden swab.

Wood-shafted cotton swabs are available at some drugstores as well as larger online drugstores.

(1) 2½-by-2½-inch square of cardboard (like manila folder cardboard)

Transparent tape

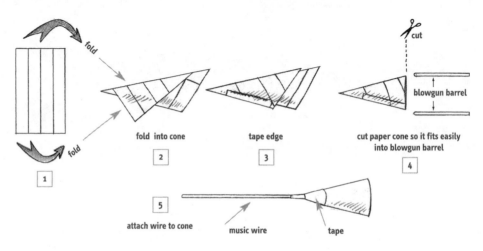

1

fold · fold

2 — fold into cone

3 — tape edge

4 — cut / blowgun barrel / cut paper cone so it fits easily into blowgun barrel

5 — attach wire to cone / music wire / tape

CONSTRUCTION

1. Roll the cardboard into an extended cone (as shown in the diagram). Wrap the cone with transparent tape to keep its shape.
2. Punch a hole in the tip of the cone that is the same diameter as the coat-hanger wire or cotton swab.
3. Insert the dart body into the hole. Tape the cone to the dart body securely, with the open end of the cone at the back of the dart.
4. Place the dart in the blowgun muzzle. The back of the dart will extend beyond the muzzle. Trace the muzzle outline on the cone using a marker. Remove the dart and cut the cone along the tracing. Now the cone should fit the inside of the blowgun exactly. You may cut the cone slightly if necessary to ensure that the dart travels freely within the blowgun.
5. The quality of the dart is very important. The cone should be squarely attached to the wire or cotton swab and centered on it exactly. The cone must fit the inside of the barrel as closely as possible without sticking.

USING THE BLOWGUN

1. Look down the barrel to make certain it is clear and free from all obstructions.
2. Unscrew the mouthpiece and insert the dart. Replace the mouthpiece.
3. Turn your head away from the mouthpiece and take a deep breath. Then, while holding your breath, bring the mouthpiece to your lips. Blow a quick hard burst of air into the blowgun. Never inhale through the blowgun.
4. Aiming takes practice. Most blowgun users leave both eyes open while they aim and shoot.

ALEXANDER VON HUMBOLDT AND THE FLYING DEATH

The small diameter and relative lightness of blowgun darts mean that the projectile must be somehow augmented to be useful to a hunter. No animal larger than a mouse could be hunted effectively by blowgun unless the lethality of the dart is increased. The most obvious way of doing so is to use poison darts.

To be useful, the poison must have two important qualities. First, it must be fast-acting, for no hunter wants to track a wounded animal all day, waiting for it to finally keel over. Secondly, the meat from the animal must remain safe to eat. Eighteenth-century Europeans knew that native hunters in South America used such poisons, but the composition of the toxin was not known until the turn of the nineteenth century, when Alexander von Humboldt solved the mystery.

In 1803, while the great American explorers Meriwether Lewis and William Clark were mapping the American Northwest, another great explorer was doing the same thing about four thousand miles to the south. A Prussian baron, assisted by a French doctor, had set out on a grand and ambitious mission to explore previously little-known areas of Central and South America.

Burning with a passion for increasing his personal store of knowledge, Baron Alexander von Humboldt, assisted by his friend Dr. Aimé Bonpland, left La Coruña, Spain, on June 5, 1799, on a ship bound for the New World. They spent the next five years investigating, exploring, cataloging, and documenting the geography, geology, flora, and fauna of Cuba, Venezuela, New Granada (now called Colombia and Ecuador), Peru, and Mexico.

The influence of the Humboldt expedition's discoveries on the sciences of biology and geography is hard to overestimate. Humboldt and Bonpland collected a vast number of plants and animals. They took rock samples while they mapped and charted. They paddled up the Magdalena River in Colombia, then continued on foot across the rugged Andes to Lima, Peru. Humboldt and Bonpland attempted to summit what was at the time the highest Western-known mountain on earth. (Everest was not yet known to non-Himalayans.) Humboldt climbed to within fourteen hundred feet of the summit of present-day Ecuador's highest peak, Chimborazo, but was stopped by an impassable vertical wall. Still, the expedition reached an elevation of nearly 22,000 feet. For the next thirty-six years Humboldt and Bonpland held the Western record for attaining the greatest height above sea level.

Humboldt and company were the first Europeans to see an electric eel, to map the ruins of ancient Incan cities, and to verify the existence of a small but important body of water called the Casiquiare Canal. The Casiquiare Canal is the only natural canal in the world that connects two major rivers, the Orinoco River and the Negro River.

The party wended its way eighteen hundred miles up the Orinoco River, charting it as they went. Perhaps equally important was Humboldt's work measuring the speed and direction of the ocean current off the coast of Peru, now called the Humboldt Current. A cold ocean current that runs along much of the western coast of South America, the Humboldt Current has tremendous effects on the ecology, weather, and economy of the entire Pacific coast.

No doubt about it, this trip was very difficult. At times the explorers had no food, and the remote, humid jungle conditions were almost unbearable. But the knowledge they gained was more than worth the hardship. Arguably, the best-known single episode during the trip occurred when Baron von Humboldt became the first European to witness South American Indians extracting and preparing a poisonous compound called curare. Up to then curare's composition was a secret known only to the inhabitants of Amazonia. For as long as the tribes could remember, they had smeared the tips of their blowgun darts with curare.

Known as the Flying Death, curare's action is swift and deadly. A simple prick from a curare dart is enough to cause death within minutes. So powerful is the toxin that it can kill a small animal in seconds, a monkey in five minutes, and a grown man in ten.

In his account of the experience, Humboldt wrote that the natives used the Flying Death for "war, the chase and gastric obstruction." Now, that information wasn't new, for references to the drug go as far back as the voyages of Sir Walter Raleigh, and other explorers had posted accounts of the legendary South American poison.

What was new was that Humboldt identified the plants from which curare was made and described what was done with them. The Indians of the area, called the Ticunas, showed him the curare plant, which modern scientists refer to as *Strychnos toxifera*, its Latin name. He also found that the poison could be safely transported in bamboo tubes, that it could be tasted without danger, and when so used was an excellent aid for stomach trouble.

He watched and recorded as the Ticunas soaked the curare plant's roots, removed its bark with a knife or machete, and pressed out the juice with their hands. Mixed with some other compounds for handling, the juicy extract was carefully dried. While hunting, the natives would take small arrows or darts and dip the sharp tips in the curare.

Curare is a very powerful and dangerous poison. The hunters used a long, hollow blowgun to shoot a single, tiny, deadly dart at prey. Very often the target was a monkey. The potency of the Flying Death varied. If the monkey, after being shot with the poison dart, fell down after hopping from one tree to a single other, the Ticunas proclaimed it "one-tree" curare. If the monkey made it to a second tree before taking its fatal plunge, it was "two-tree." And if the curare wasn't quite up to par, it was "three-tree" poison, which entailed a lot of running around by the hunter to find and gather the paralyzed prey.

The curare Humboldt brought back was carefully analyzed. Since that time the drug has been used extensively in Western medicine, and it continues to be used in a variety of medical applications, such as muscle relaxants and anesthetics.

After returning home, Humboldt led a long and active life. He wrote extensively of his expeditions and discoveries. One of his books, *A Personal Narrative*, was important to young Charles Darwin and may have inspired him to take up his studies in biology.

When he was sixty, Humboldt traveled to Siberia and Central Asia to study the weather. He died at age ninety and was buried in Tegel, Germany. Today Humboldt is widely remembered as a great explorer and geographer. Fittingly, several landmarks in the Americas bear his name, including the Humboldt Current, river, and mountain range. The Mare Humboldtianum, a large area on the northeastern face of the moon, was named for him.

Conclusion

There are two people whose lives and works really encapsulate the culture of those who experiment with the projects and histories described here. Neither is very well known outside their small group of admirers, but they're both definitely worth knowing.

The first is Saint Barbara. Barbara is the patron saint of artillerymen, fireworks makers, bomb disposal technicians, miners, and anybody else whose actions put them in contact with things that go boom.

Most of the stories surrounding the life of Saint Barbara are myths that have arisen out of earlier legends and folklore. According to one of many legends involving Saint Babs (as artillerymen who look to her for protection often call her), she was a young and pious woman who lived in the area of Nicomedia in fifth-century Asia Minor. In the summer of 430 an army of Vandals marching eastward from their Germanic homelands arrived at the walls of the city. The Nicomedians took one look at the size of the approaching army and quickly barricaded themselves behind the high, thick town walls. A long siege began.

The leader of the city, a man named Alypius, retrieved his daughter Barbara from the convent where she lived, to assist in the defense of the city. Alypius and Barbara took large jars and filled them with a mysterious substance known only to them. They ordered that the jars' contents be poured into the trenches surrounding the city, which then ignited into a towering inferno, an impenetrable wall of flame that repelled the attackers.

After a pause to regroup and reconsider, the Vandals continued their siege. Over the next year Barbara and her charges repeatedly repelled the attackers by using a variety of incendiary weaponry, including, so the legend says, burning globes of fire hurled from catapults.

In the end, though, the numerical superiority of the besiegers overcame the city's defenses. The besieging army stormed into the convent where Barbara had sought refuge. But the warrior-saint had a final trick up her sleeve. She had amassed a huge store of the mysterious explosive in a subterranean passage beneath the convent.

On Barbara's order, a gigantic explosion was set off, and the fireball consumed both the Vandal horde and Barbara; the Vandals were trapped beneath the debris of the exploded convent, and Barbara shot heavenward presumably to glory and sainthood.

Besides having their own patron saint, those who work with things that go boom have a patron philosopher as well. Georg Lichtenberg was a distinguished eighteenth-century German scientist and scholar. Lichtenberg, while somewhat well known today, was an intellectual luminary of his time. He was a fellow of Britain's Royal Society, an esteemed professor at Göttingen University, and a friend and confidant of such leading lights as Immanuel Kant, Johann Wolfgang von Goethe, and Alessandro Volta. And he loved fast-moving, noisy things.

The polymathic Lichtenberg was an experimental physicist, astronomer, mathematician, and diarist. He is given credit for being among the first university instructors to incorporate experimentation into his classes, and he is the discoverer of xerography, from which modern copying machines descended.

But arguably, he is best remembered for his quotable aphorisms, all of which are short, clever, and razor sharp.

One of Lichtenberg's best-known quotes is a favorite of mine and makes for a fitting conclusion to this book. So as you attempt to build the projects described in the previous pages, take a moment to remember what Lichtenberg wrote regarding loud, kinetic, and energetic experiments.

"A physical experiment which makes a bang," he wrote, "is always worth more than a quiet one. Therefore a man cannot strongly enough ask of Heaven: if it wants to let him discover something, may it be something that makes a bang. It will resound into eternity."

Acknowledgments

This book is the result of scopious input from many people. Thanks to the many readers of my past books who took the time to write me about their experiences and to share ideas. Please feel free to send me ideas by email to whoosh@gurstelle.com. While I can't always reply promptly, I always appreciate them.

My sincere gratitude goes to those who provided inspiration, project suggestions, and candid advice. Steve Cox, Jim Keifenheim, Jerry Pohlen, Caz Sienkiewicz, and my many friends at *Make Magazine* are first among them.

As always, my capable agent, Jane Dystel, provided great advice and service. Thanks to the many people at Crown Books, including Elissa Altman, Aliza Fogelson, and Nikki Van Noy, who made sure that nonexistent words such as *scopious* never made it into the final copy.

Index